Eyal Kolman and Michael Margaliot

Knowledge-Based Neurocomputing: A Fuzzy Logic Approach

Studies in Fuzziness and Soft Computing, Volume 234

Editor-in-Chief

Prof. Janusz Kacprzyk
Systems Research Institute
Polish Academy of Sciences
ul. Newelska 6
01-447 Warsaw
Poland
E-mail: kacprzyk@ibspan.waw.pl

Further volumes of this series can be found on our homepage: springer.com

Vol. 220. Humberto Bustince,
Francisco Herrera, Javier Montero (Eds.)
*Fuzzy Sets and Their Extensions:
Representation, Aggregation and Models*, 2007
ISBN 978-3-540-73722-3

Vol. 221. Gleb Beliakov, Tomasa Calvo,
Ana Pradera
*Aggregation Functions: A Guide
for Practitioners*, 2007
ISBN 978-3-540-73720-9

Vol. 222. James J. Buckley,
Leonard J. Jowers
*Monte Carlo Methods in Fuzzy
Optimization*, 2008
ISBN 978-3-540-76289-8

Vol. 223. Oscar Castillo, Patricia Melin
*Type-2 Fuzzy Logic: Theory and
Applications*, 2008
ISBN 978-3-540-76283-6

Vol. 224. Rafael Bello, Rafael Falcón,
Witold Pedrycz, Janusz Kacprzyk (Eds.)
*Contributions to Fuzzy and Rough Sets
Theories and Their Applications*, 2008
ISBN 978-3-540-76972-9

Vol. 225. Terry D. Clark, Jennifer M. Larson,
John N. Mordeson, Joshua D. Potter,
Mark J. Wierman
*Applying Fuzzy Mathematics to Formal
Models in Comparative Politics*, 2008
ISBN 978-3-540-77460-0

Vol. 226. Bhanu Prasad (Ed.)
Soft Computing Applications in Industry, 2008
ISBN 978-3-540-77464-8

Vol. 227. Eugene Roventa, Tiberiu Spircu
*Management of Knowledge Imperfection in
Building Intelligent Systems*, 2008
ISBN 978-3-540-77462-4

Vol. 228. Adam Kasperski
Discrete Optimization with Interval Data, 2008
ISBN 978-3-540-78483-8

Vol. 229. Sadaaki Miyamoto,
Hidetomo Ichihashi, Katsuhiro Honda
Algorithms for Fuzzy Clustering, 2008
ISBN 978-3-540-78736-5

Vol. 230. Bhanu Prasad (Ed.)
Soft Computing Applications in Business, 2008
ISBN 978-3-540-79004-4

Vol. 231. Michal Baczynski,
Balasubramaniam Jayaram
Soft Fuzzy Implications, 2008
ISBN 978-3-540-69080-1

Vol. 232. Eduardo Massad,
Neli Regina Siqueira Ortega,
Laécio Carvalho de Barros,
Claudio José Struchiner
*Fuzzy Logic in Action: Applications
in Epidemiology and Beyond*, 2008
ISBN 978-3-540-69092-4

Vol. 233. Cengiz Kahraman (Ed.)
*Fuzzy Engineering Economics with
Applications*, 2008
ISBN 978-3-540-70809-4

Vol. 234. Eyal Kolman, Michael Margaliot
*Knowledge-Based Neurocomputing:
A Fuzzy Logic Approach*, 2009
ISBN 978-3-540-88076-9

Eyal Kolman and Michael Margaliot

Knowledge-Based Neurocomputing: A Fuzzy Logic Approach

 Springer

Authors

Eyal Kolman
School of Electrical Engineering
Tel Aviv University
69978 Tel Aviv
Israel
E-mail: eyal@eng.tau.ac.il

Michael Margaliot, Ph.D.
School of Electrical Engineering
Tel Aviv University
69978 Tel Aviv
Israel
E-mail: michaelm@eng.tau.ac.il

ISBN 978-3-540-88076-9   e-ISBN 978-3-540-88077-6

DOI 10.1007/978-3-540-88077-6

Studies in Fuzziness and Soft Computing   ISSN 1434-9922

Library of Congress Control Number: 2008935494

© 2009 Springer-Verlag Berlin Heidelberg

This work is subject to copyright. All rights are reserved, whether the whole or part of the material is concerned, specifically the rights of translation, reprinting, reuse of illustrations, recitation, broadcasting, reproduction on microfilm or in any other way, and storage in data banks. Duplication of this publication or parts thereof is permitted only under the provisions of the German Copyright Law of September 9, 1965, in its current version, and permission for use must always be obtained from Springer. Violations are liable to prosecution under the German Copyright Law.

The use of general descriptive names, registered names, trademarks, etc. in this publication does not imply, even in the absence of a specific statement, that such names are exempt from the relevant protective laws and regulations and therefore free for general use.

Typeset & Cover Design: Scientific Publishing Services Pvt. Ltd., Chennai, India.

Printed in acid-free paper

9 8 7 6 5 4 3 2 1

springer.com

To my parents and my wife. All that I am, I owe to you – E.K.

To three generations of women: Hannah, Miri, and Hila – M.M.

I have set before you life and death,
the blessing and the curse.
So choose life in order that you may live,
you and your descendants. (Deuteronomy 30:19)

Preface

Artificial neural networks (ANNs) serve as powerful computational tools in a diversity of applications including: classification, pattern recognition, function approximation, and the modeling of biological neural networks. Equipped with procedures for learning from examples, ANNs can solve problems for which no algorithmic solution is known.

A major shortcoming of ANNs, however, is that the knowledge learned by the network is represented in an exceedingly opaque form, namely, as a list of numerical coefficients. This *black-box* character of ANNs hinders the possibility of more widespread acceptance of them, and makes them less suitable for medical and safety-critical applications.

A very different form of knowledge representation is provided by *fuzzy rule-bases* (FRBs). These include a collection of If-Then rules, stated in *natural language*. Thus, the knowledge is represented in a form that humans can understand, verify, and refine. In many cases, FRBs are derived based on questioning a human expert about the functioning of a given system. Transforming this information into a complete and consistent set of rules, and determining suitable parameter values, is a nontrivial challenge.

It is natural to seek a synergy between the plasticity and learning abilities of ANNs and the transparency of FRBs. Indeed, considerable research attention has been devoted to the development of various neuro-fuzzy models, but this synergy is a target yet to be accomplished.

In this monograph, we introduce a novel FRB, referred to as the *Fuzzy All-permutations Rule-Base* (FARB). We show that inferring the FARB, using standard tools from fuzzy logic theory, yields an input-output relationship that is *mathematically equivalent* to that of an ANN. Conversely, every standard ANN has an equivalent FARB. We provide the explicit bidirectional transformation between the ANN and the corresponding FARB.

The FARB–ANN equivalence integrates the merits of symbolic FRBs and subsymbolic ANNs. We demonstrate this by using it to design a new approach for knowledge-based neurocomputing using the FARB. First, by generating the equivalent FARB for a given (trained) ANN, we immediately obtain a symbolic representation of the knowledge learned by the network. This provides a novel and simple method for knowledge extraction from trained ANNs.

VIII   Preface

The interpretability of the FARB might be hampered by the existence of a large number of rules or complicated ones. In order to overcome this, we also present a systematic procedure for rule reduction and simplification. We demonstrate the usefulness of this approach by applying it to extract knowledge from ANNs trained to solve: the Iris classification problem, the LED display recognition problem, and a formal language recognition problem.

Second, stating initial knowledge in some problem domain as a FARB immediately yields an equivalent ANN. This provides a novel approach for knowledge-based design of ANNs. We demonstrate this by designing recurrent ANNs that solve formal language recognition problems including: the AB language, the balanced parentheses language, and the 0^n1^n language. Note that these languages are context-free, but not regular, so standard methods for designing RNNs are not applicable in these cases.

Some of the results described in this work appeared in [89, 88, 90, 91]. We are grateful to several anonymous reviewers of these papers for providing us with useful and constructive comments.

We gratefully acknowledge the financial support of the Israeli Ministry of Science and Technology, the Israel Science Foundation, and the Adams Super Center for Brain Research at Tel Aviv University.

Tel Aviv, Israel   Eyal Kolman
July 2008   Michael Margaliot

Abbreviations

AI	Artificial Intelligence
ANN	Artificial Neural Network
COG	Center Of Gravity
DFA	Discrete Finite-state Automaton
DOF	Degree Of Firing
FARB	Fuzzy All-permutations Rule-Base
FFA	Fuzzy Finite-state Automaton
FL	Fuzzy Logic
FRB	Fuzzy Rule-Base
IFF	If and only if
IO	Input-Output
KBD	Knowledge-Based Design
KBN	Knowledge-Based Neurocomputing
KE	Knowledge Extraction
LED	Light Emitting Diode
MF	Membership Function
MLP	Multi-Layer Perceptron
MOM	Mean Of Maxima
RBFN	Radial Basis Function Network
RNN	Recurrent Neural Network
SVM	Support Vectors Machine

Symbols

$\mathbf{x}$	column vector
$\mathbf{x}^T$	transpose of vector $\mathbf{x}$
x_i	ith element of the vector $\mathbf{x}$
$||\mathbf{x}||$	Euclidean norm of the vector $\mathbf{x}$
$[j:k]$	set of integers $\{j, j+1, \ldots, k\}$
$\mathbb{R}^n$	space of n-dimensional real-valued numbers
$\mu_{term}(\cdot)$	membership function for the fuzzy set $term$
i-or	interactive-or operator (see p. 8)
$\sigma(\cdot)$	logistic function (see p. 8)
$\sigma_L(\cdot)$	piecewise linear logistic function (see p. 18)
$h^{-1}(\cdot)$	inverse of the function $h(\cdot)$
$\&$	logical and operator
L_4	language generated by Tomita's 4th grammar
$t^k(\mathbf{x})$	truth value of Rule k for input $\mathbf{x}$
$Prob(A)$	probability of event A
$E\{x\}$	expected value of the random variable x

Contents

Preface VII

List of Abbreviations IX

List of Symbols XI

1   Introduction 1
   1.1 Artificial Neural Networks (ANNs) 2
   1.2 Fuzzy Rule-Bases (FRBs) 3
   1.3 The ANN–FRB Synergy 4
   1.4 Knowledge-Based Neurocomputing 5
      1.4.1 Knowledge Extraction from ANNs 5
      1.4.2 Knowledge-Based Design of ANNs 9
   1.5 The FARB: A Neuro-fuzzy Equivalence 11

2   The FARB 13
   2.1 Definition 15
   2.2 Input-Output Mapping 18

3   The FARB–ANN Equivalence 21
   3.1 The FARB and Feedforward ANNs 21
      3.1.1 Example 1: Knowledge Extraction from a
            Feedforward ANN 22
      3.1.2 Example 2: Knowledge-Based Design of a
            Feedforward ANN 24
   3.2 The FARB and First-Order RNNs 26
      3.2.1 First Approach 26
      3.2.2 Example 3: Knowledge Extraction from a
            Simple RNN 27
      3.2.3 Second Approach 28

XIV   Contents

        3.2.4   Third Approach 29
        3.2.5   Example 4: Knowledge Extraction from an RNN 30
        3.2.6   Example 5: Knowledge-Based Design of an RNN 32
   3.3   The FARB and Second-Order RNNs 33
   3.4   Summary 35

4   Rule Simplification 37
   4.1   Sensitivity Analysis 37
   4.2   A Procedure for Simplifying a FARB 39

5   Knowledge Extraction Using the FARB 41
   5.1   The Iris Classification Problem 41
   5.2   The LED Display Recognition Problem 44
        5.2.1   Knowledge Extraction Using the FARB 46
        5.2.2   FARB Simplification 46
        5.2.3   Analysis of the FRB 48
   5.3   The L_4 Language Recognition Problem 50
        5.3.1   Formal Languages 50
        5.3.2   Formal Languages and RNNs 51
        5.3.3   The Trained RNN 51
        5.3.4   Knowledge Extraction Using the FARB 53

6   Knowledge-Based Design of ANNs 59
   6.1   The Direct Approach 60
        6.1.1   KBD of an ANN Recognizing L_4 60
   6.2   The Modular Approach 63
        6.2.1   The Counter Module 63
        6.2.2   The Sequence-Counter Module 66
        6.2.3   The String-Comparator Module 66
        6.2.4   The String-to-Num Converter Module 67
        6.2.5   The Num-to-String Converter Module 68
        6.2.6   The Soft Threshold Module 68
        6.2.7   KBD of an RNN for Recognizing the Extended
              L_4 Language 69
        6.2.8   KBD of an RNN for Recognizing the AB Language 71
        6.2.9   KBD of an RNN for Recognizing the Balanced
              Parentheses Language 72
        6.2.10 KBD of an RNN for Recognizing the 0^n1^n Language 74

7   Conclusions and Future Research 77
   7.1   Future Research 77
        7.1.1   Regularization of Network Training 78
        7.1.2   Extracting Knowledge during the Learning Process 79
        7.1.3   Knowledge Extraction from Support Vector Machines ... 79
        7.1.4   Knowledge Extraction from Trained Networks 80

A	Proofs	83
B	Details of the LED Recognition Network	87
References		89
Index		99

1 Introduction

In this chapter, we introduce some of the basic themes of this work. We begin with some background material on artificial intelligence and the paradigms of connectionism and symbolism. This is followed by a very brief review of artificial neural networks (ANNs) and fuzzy rule-based systems (FRBs). This sets the stage for the presentation of the main tool developed in this work, the *fuzzy all-permutations rule-base* (FARB). This is an FRB with a special form. This form guarantees that the input-output mapping of the FARB is *mathematically equivalent* to that of an ANN. This provides a new synergy between the learning-from-examples ability of ANNs and the high-level symbolic information processing of FRBs.

The modern field of *artificial intelligence* (AI) [147] was born at a conference held at the campus of Dartmouth College in the summer of 1956 [109]. John McCarthy, who coined the term AI, defined it as "the science and engineering of making intelligent machines"[108].

A natural approach for developing such intelligent machines is based on an attempt to mimic the human reasoning process or the behavior of some ideal rational agent.[1] Two paradigms for explaining and imitating the human reasoning process are *symbolism* and *connectionism* [40, 161, 179]. Symbolism views reasoning as the process of creating and manipulating a symbolic map of the outside world. This approach is based on using a formal axiom system for symbol processing. Axioms, theorems, and deductive rules are used to manipulate symbols in order to derive meaningful conclusions. Examples of successful applications of symbolism in AI include the use of rule-based systems that imitate the reasoning process of human experts in some specific, and usually narrow, domain [70, 147, 159].

The connectionist approach is inspired by biological neural networks. The human brain information processing ability is thought to emerge primarily from the interactions of large interconnected networks of neurons. *Artificial neural networks* (ANNs) are based on an attempt to imitate this idea. The development of suitable training algorithms provided ANNs with the ability to learn

---

[1] Human decision making and rationality are not always congruent [79].

and generalize from examples. In this respect, ANNs provide a solution to the problem posed by Arthur Samuel [151] in 1959:

> "How can computers be made to do what needs to be done, without being told exactly how to do it?"

1.1 Artificial Neural Networks

In 1943, McCulloch and Pitts proposed a model for an artificial neural network [110]. Motivated by the theory of finite-state machines, each neuron in the network had two possible states: ON and OFF. The state is determined according to the neuron inputs, which are themselves the outputs of neighboring neurons (see Fig. 1.1). McCulloch and Pitts showed how such an ANN can perform simple logical functions, and suggested that ANNs may have learning capabilities.

In 1949, Hebb suggested a simple updating rule for modifying the network connections [60]: if two neurons are activated at the same time, then their connection-weight is increased; otherwise, it is decreased. This so-called *Hebbian learning* remains an influential idea in the field of machine learning till today.

In 1957, Rosenblatt presented the *perceptron*, which is a linear classifier implemented by a single neuron [141]. A major result was the presentation of a simple and converging learning algorithm for this classifier. About a decade later, Minsky and Papert provided a very influential critical overview of the perceptron [116].

In 1982, Hopfield used techniques from statistical mechanics to analyze the storage and optimization properties of dynamic ANNs, treating the neurons as a collection of atoms. A major breakthrough occurred in the mid-1980s, when several groups reinvented the back-propagation algorithm [95, 130, 144, 180], a highly efficient training algorithm for multi-layer perceptrons (MLPs).

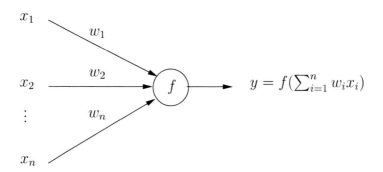

Fig. 1.1. An artificial neuron: $x_1, \ldots, x_n$ are the inputs, w_i are the neuron weights, and y is the output

ANNs proved to be a highly successful distributed computation paradigm, and were effectively used to solve tasks in the fields of classification, pattern recognition, function approximation, and more. Many results were gathered in the influential collection [145].

The ability of ANNs to learn and generalize from examples makes them very suitable for many real-world applications where exact algorithmic approaches are either unknown or too difficult to implement. Numerous problems that are hard to address using conventional algorithms were solved by training MLPs using the backprop algorithm. ANNs were also extensively used in the modeling and analysis of biological neural networks [14, 29, 59, 115, 192].

However, the knowledge learned during the training process is embedded in a complex, distributed, and sometimes self-contradictory form [69]. Whether the ANN operates properly or not, it is very difficult to comprehend exactly what it is computing. In this respect, ANNs process information on a "black-box" and sub-symbolic level.

A very different form of knowledge representation is provided by fuzzy rule-based systems.

1.2 Fuzzy Rule-Bases

Rule-based systems play an important role in the symbolic approach to AI. The system knowledge is stated as a collection of If-Then rules. Inferring the rules provides the system input-output mapping. A typical application of rule-based systems is the design of *expert systems* that mimic the reasoning process of a human expert, in a certain, and usually narrow, knowledge domain. Expert systems are built by restating the knowledge of the human expert as a suitable set of rules. Two famous examples are the DENDRAL system, which was used to infer the molecular structures of materials from mass spectrometer data [21]; and the MYCIN medical diagnostic system. The performance of MYCIN was comparable to that of human physicians [159].

A major difficulty in the design of expert systems is that human experts explain their reasoning process using natural language which is inherently vague and imprecise. Transforming this information into logical If-Then rules is a highly nontrivial task.

In the mid 1960s, Lotfi A. Zadeh pointed out that classical logic theory is not always appropriate in the context of human reasoning. His *principle of incompatibility* states that if a system is complex enough, precision and relevance (or meaningfulness) of statements about its behavior are mutually exclusive characteristics [185]. To overcome these difficulties, Zadeh introduced *fuzzy sets* and *fuzzy logic* (FL). Fuzzy sets [184] are a generalization of ordinary sets. The membership of an element in a fuzzy set is not binary, but rather it is a continuum grade in the range $[0, 1]$. FL is an extension of classical logic theory based on using fuzzy sets and fuzzy logic operators. This provides the machinery needed to handle propositions whose truth values are not necessarily binary.

4   Introduction

Zadeh pointed out that FL theory is suitable for modeling human language and the human reasoning process. More recently, Zadeh suggested the notion of *computing with words* [187], as opposed to the classic paradigm of computation with exact numerical values and symbols. Indeed, it soon became evident that the true power of fuzzy logic lies in its ability to handle and manipulate *linguistic information* [33, 107, 174, 186, 187].

In particular, *fuzzy rule-bases* (FRBs) include a collection of If-Then rules, stated using *natural language*, and describing how the input affects the output. Thus, the knowledge is expressed in a form that humans can easily understand, verify, and refine. In this respect, FRBs are also useful for expressing partial or self-contradicting knowledge due to their ability to handle vagueness and uncertainty [122].

In 1974, Mamdani and his colleagues [103, 104] designed a *fuzzy controller* that regulated a steam engine. The engine had two inputs and two outputs. The inputs were the heat temperature to the boiler and throttle opening of the engine cylinder. The regulated outputs were the steam pressure and the engine speed. The fuzzy controller was an FRB composed of rules in the form:

If the outputs are ... Then the inputs should be ....

These rules were derived using common sense and stated using natural language. An example is the rule:

If speed error is *positive big*
    and change in speed error is *not(negative big* or *negative medium)*
Then change in the throttle opening must be *negative big*.

This first practical application of fuzzy rule-based systems proved to be a startling success, and quickly led the way to the application of FRBs in numerous real-world applications, ranging from household electrical devices [72] to subway trains [165].

In summary, FRBs provide efficient means for transforming verbally-stated knowledge into an algorithmic form. Thus, FRBs are frequently used to develop expert systems [54, 164, 165], as well as to transform verbal descriptions of various phenomena into well-defined mathematical models [136, 143, 174, 175]. However, a major disadvantage of FRBs is the lack of a systematic algorithm for determining the fuzzy sets and the fuzzy operators suitable for a given problem. A natural idea is to integrate some kind of learning capability into the FRB.

1.3   The ANN–FRB Synergy

A great deal of research has been devoted to the design of hybrid intelligent systems that fuse sub-symbolic and symbolic techniques for information processing [80, 111], and, in particular, to creating an ANN–FRB synergy [118], or neuro-fuzzy models. The desired goal is a combined system demonstrating the robustness and learning capabilities of ANNs and the "white-box"[2] character of FRBs.

---

[2] "White-box" is the antonym of "black-box", i.e., a system with behavior and conclusions that can be explained and analyzed in a comprehensible manner.

Fuzzy logic techniques have been used for improving ANN features, by using fuzzy preprocessing of the training data, fuzzy clustering of the data, and fuzzy learning algorithms [132]. Particular attention was devoted to special ANN architectures that perform fuzzy information processing [133]. Examples include the famous *Adaptive Network-based Fuzzy Inference System* (ANFIS) [73, 75], which is a feedforward network representation of the fuzzy reasoning process,[3] and Lin and Lee's *Neural-Network-Based Fuzzy Logic Control and Decision System* [98], which is a neuro-implementation of a fuzzy controller. In both cases, nodes in different layers perform different tasks corresponding to the different stages in the fuzzy reasoning process.[4]

Another form of a neuro-fuzzy synergy that has been extensively discussed in the literature is based on transforming a given ANN into a corresponding fuzzy rule-base, and vice versa (see the surveys [4, 71, 118, 166]).

1.4 Knowledge-Based Neurocomputing

Knowledge-based neurocomputing (KBN) concerns the use and representation of *symbolic* knowledge within the neurocompting paradigm. The focus of KBN is therefore on methods to encode prior knowledge, and to extract, refine, and revise knowledge embedded within an ANN [29]. Cloete sorted KBN techniques into three types: *unified*, *hybrid*, and *translational* [27]. In unified architectures, no explicit symbolic functionality is programmed into the ANN. In hybrid architectures, separate ANNs are responsible for particular functions and knowledge is transferred between them. Translational architectures maintain two knowledge representations: subsymbolic and symbolic, and the information flows back and forth between them.

Extracting and representing the knowledge that was learned by a trained ANN in the form of symbolic rules is referred to as *knowledge extraction* (KE), whereas designing ANNs on the basis of prior knowledge is called *knowledge insertion* or *knowledge-based design* (KBD).

1.4.1 Knowledge Extraction from ANNs

The main drawback of ANNs is their black-box character. The knowledge embedded in the ANN is distributed in the weights and biases of the different neurons, and it is very difficult to comprehend exactly what the ANN is computing. This hinders the possibility of more widespread acceptance of ANNs, and makes them less suitable for certain applications. For example, in many medical applications black-box decisions are deemed unacceptable [177]. Specifically, the *learned intermediaries doctrine* places on the clinician a responsibility to understand any inferences derived from an assisting model [38].

---

[3] This is a specific implementation of logic-based neural networks [131], where neurons apply logical functions.

[4] For example, the ANFIS first layer nodes compute membership function values, whereas nodes in the second layer perform T-norm operations.

6 Introduction

The importance of the KE challenge cannot be overemphasized. Several potential benefits include [30, 32]:

- **Validation.** Explaining the ANNs functioning in a comprehensible form provides a validation of its suitability for the given task. This is vital in safety-critical applications [19] [31].
- **Feature extraction.** During the training process, ANNs learn to recognize the relevant features in enormous sets of data. Understanding what these features are may lead to a deeper theoretical understanding of the problem at hand. Efficient feature extraction may also lead to an improvement in the overall accuracy of the system, and to efficient compression schemes [152].
- **Refinement and improvement.** A comprehensible model can be analyzed and refined. This can increase precision, reduce inconsistencies, and improve the overall ability to generalize and predict correctly.
- **Scientific discovery.** ANNs trained using a set of classified examples can learn to solve difficult classification problems for which no algorithmic approach is known. Understanding the functioning of these ANNs can make them accessible for human review and lead to new discoveries [67].

Currently, all existing KE techniques suffer serious drawbacks, and the problem is far from being resolved. The overwhelming implications of successful KE have made this topic the "holy grail" of research in the field. We present a brief overview of existing KE techniques. For more details, see [4, 29, 166].

Knowledge Extraction from Feedforward ANNs

KE techniques can be sorted into one of three types. The *decompositional* methodology [4, 167] generates rules for each hidden or output neuron, by searching for combinations of inputs whose weighted sum exceeds the neuron bias. Then, rules are created with the discovered combination as a premise. Examples include Fu's *KT (Knowledge Translator)* [43] and Towell and Shavlik's *Subset* [172] algorithms. A variant is the *MofN* method, that generates rules of the form *"If at least M of the following N premises are true, Then ..."* [172].

The *pedagogical* approach [4] treats the ANN as a black-box and generates a knowledge representation that has the same (or similar) input-output (IO) mapping, disregarding the specific architecture of the network. Input-output pairs are generated using the trained network, and rules are extracted from this new database. Examples include Craven's *TREPAN (TREes PArroting Networks)* decision tree [30]; dividing the input and output domains to polyhedral regions and extracting rules that map input polyhedra to output polyhedra [61, 102]; and generating rules that connect inputs and outputs that are active at the same time [148]. Some techniques analyze the network input-output mapping by computing the derivative of the output with respect to a specific input [68], or by discrete rule extraction through orthogonal search [38].

Models that use both decompositional and pedagogical approaches, are referred to as *eclectic*. For example, the DEDEC algorithm [168] extracts rules from an ANN based on its IO mapping, but ranks their importance using the network coefficients.

Several approaches are based on searching for the "most effective" IO paths in the ANN, and extracting rules that approximate the performance of these paths [18, 55, 119, 155]. For example, Mitra and Pal presented the *Fuzzy-MLP* [117, 119], which is a feedforward network with fuzzified inputs. After training, fuzzy rules are extracted by finding the input and hidden neurons that had the largest influence on the output, for a given input.

One reason the KE problem is difficult is that the knowledge learned during training is embedded in a highly distributed and complex form. A very useful technique is modification of the training algorithm by adding suitable regularization terms to the error criterion [150]. The goal is to force the ANN to develop a more condensed and skeletal structure, which facilitates subsequent KE [14, 156]. Ishikawa [69, 114] incorporates regularization terms that punish large weights. Duch *et al.* [35, 36] use regularization terms that force the weights to one of the values $\{-1,1,0\}$. In both methods, KE is then performed by representing the IO mapping of each neuron as a Boolean function.

A common drawback of nearly all existing KE techniques is that they use *approximated representations*. Indeed, one of the common evaluation criteria for KE techniques is *fidelity* [4, 190], i.e., how well one representation mimics the other. For a KE technique yielding a fidelity error close to zero, but not equal to zero, see [102]. To the best of our knowledge, there are only two algorithms that generate representations with a zero fidelity error. We now discuss them in more detail.

The Jang and Sun model

Jang and Sun [74] noted that the activation functions of radial basis function networks (RBFNs) are the Gaussian membership functions frequently used in FRBs. They used this to extract a FRB that is mathematically equivalent to the RBFN. In other words, the IO mapping of the FRB is mathematically equivalent to the IO mapping of the original RBFN. Yet, this equivalence holds only for RBFNs, and each membership function can be used by no more than one rule [3].

The Benitez et al. model

Benitez *et al.* [12] showed that ANNs with Logistic activation functions are mathematically equivalent to Mamdani-type FRB, but with some special and non-standard fuzzy operators. Their method is quite relevant to the approach presented in this monograph and so we review their model in detail.

Consider a feedforward ANN with k inputs z_i, $i = 1, \ldots, k$, a hidden layer of n neurons with activation function $h : \mathbb{R} \to \mathbb{R}$, and a single output unit (see Fig. 1.2). The ANN output is given by

$$f = c_0 + \sum_{j=1}^{n} c_j h(\sum_{i=1}^{k} w_{ji} z_i + b_j). \tag{1.1}$$

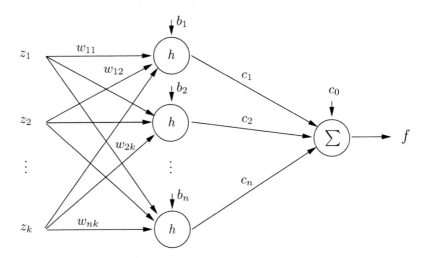

Fig. 1.2. Feedforward ANN with a single hidden layer

Let $\sigma(x)$ denote the *Logistic function*:

$$\sigma(x) := \frac{1}{1+\exp(-x)}. \tag{1.2}$$

Benitez et al. [12] (see also [24]) noted that for the particular case of an ANN with Logistic activation functions $h(x) = \sigma(x)$, Eq. (1.1) can be interpreted as the result of inferencing a set of n fuzzy rules, with rule j, $j = 1, \ldots, n$, in the form

$$R_j : \text{If} \left(\sum_{i=1}^{k} w_{ji} z_i + b_j \right) \text{ is } A \text{ Then } f = c_j. \tag{1.3}$$

The linguistic term *is A* is defined using a Logistic membership function, and the actual output is computed using a weighted (but not normalized) sum of the Then-parts of the rules.

To further simplify the If-part in the rules, Benitez et al. introduced the *interactive-or* (i-or) operator $* : \mathbb{R} \times \mathbb{R} \cdots \times \mathbb{R} \to \mathbb{R}$, defined by

$$a_1 * a_2 * \cdots * a_n := \frac{a_1 a_2 \ldots a_n}{a_1 a_2 \ldots a_n + (1-a_1)(1-a_2)\ldots(1-a_n)}. \tag{1.4}$$

Note that if at least one the a_is is zero then $a_1 * a_2 * \cdots * a_n = 0$ and that $1 * 1 * \cdots * 1 = 1$.

It is easy to verify using (1.2) that i-or is the σ-dual of the $+$ operation, that is, $\sigma(\sum_{i=1}^{n} a_i) = \sigma(a_1) * \sigma(a_2) * \cdots * \sigma(a_n)$, so (1.3) can be rewritten as

If $(w_{j1}z_1$ is $A)*(w_{j2}z_2$ is $A)* \ldots *(w_{jk}z_k$ is $A)*(b_j$ is $A)$ Then $f = c_j$. (1.5)

Thus, the If-part in rule R_j is a composition, using the i-or operator, of the $k+1$ atoms: $(w_{ji}z_i$ is $A)$, $i = 1, \ldots, k$, and $(b_j$ is $A)$.

This model yields an equivalence between an ANN with Logistic activation functions and an FRB with rules in the form (1.5). Benitez *et al.* demonstrated how this can be used to extract the knowledge embedded in a trained ANN (with n hidden neurons) in the form of an FRB (with n rules). However, the use of the i-or operator and the additive inferencing imply that this FRB is *not* a standard FRB.

More recently, extensions of Benitez' model to Takagi-Sugeno-type fuzzy systems also appeared [24, 189]. As noted in [118], models using Takagi-Sugeno-type rules are more difficult to interpret than Mamdani-type models.

Knowledge Extraction from Recurrent ANNs

The existence of feedback connections makes recurrent neural networks (RNNs) more powerful than feedforward ANNs, but also more complicated to train and analyze. RNNs are widely used in various domains, including financial forecasting [49, 93], control [112], speech recognition [138], visual pattern recognition [96], and more.

Knowledge extraction from RNNs is more difficult due to the intricate feedback connections [71]. Consequently, common KE techniques for RNNs are based not on rule extraction, but rather on transforming the RNN into a corresponding *deterministic finite-state automaton* (DFA). This is carried out in four steps [71]:

1. Quantization of the continuous state space of the RNN, resulting in a set of discrete locations.
2. Feeding the RNN with input patterns and generating the resulting states and outputs.
3. Construction of the corresponding DFA, based on the observed transitions.
4. Minimization of the DFA.

Variants of this approach include: quantization using equipartitioning of the state space [50, 126] and vector quantization [41, 52, 188]; generating the state and output of the DFA by sampling the state space [105, 178]; extracting stochastic state machines [169, 170]; extracting fuzzy state machines [16], and more [71].

However, this form of KE has several drawbacks. First, RNNs are continuous-valued and it is not at all obvious whether they can be suitably modeled using discrete-valued mechanisms such as DFAs [86, 87]. Second, the resulting DFA depends crucially on the quantization level. Coarse quantization may cause large inconsistencies between the RNN and the extracted DFA, while fine quantization may result in an overly large and complicated DFA. Finally, the comprehensibility of the extracted DFA is questionable. This is particularly true for DFAs with many states, as the meaning of every state/state-transition is not necessarily obvious. These disadvantages encourage the development of alternative techniques for KE from RNNs [139, 140].

1.4.2 Knowledge-Based Design of ANNs

In many real-world problems, certain prior knowledge on a suitable solution exists. Using this information in the design of an ANN for solving the problem

10   Introduction

is important because the initial architecture and parameter values of a network can have a critical effect on its functioning. If the initial network is too simple, it may not be able to solve the given problem for any set of parameters. If the network is too complicated, the training algorithm may not converge at all or may lead to overfitting. Additionally, standard training algorithms do not guarantee convergence to a global minimum, and are highly dependent on the initial values of the network parameters. Successful KBD can improve various features of the trained ANN (e.g., generalization capability) as well as reduce training times [8, 44, 117, 162, 173].

Knowledge-Based Design of Feedforward ANNs

A common KBD approach is based on mapping rule-based systems into a neural-like architecture: final hypotheses are represented using output neurons; data attributes become input neurons; and the rule strength is mapped into the weight of the corresponding connection. The rules are then modified using back-propagation learning. For example, Fu and Fu [45] used this scheme to map expert systems into ANNs, and Gallant [47] represented hypotheses that are either True, False, or Unknown by constraining the neuron values to $\{1,-1,0\}$, respectively.

Fuzzy rule-bases are often represented in the form of a feedforward network. The first layer fuzzifies the input. In the second layer, each neuron implements a fuzzy rule. The rules are aggregated in the third layer, and defuzzification is performed in the output layer. Examples include Towell and Shavlik's *Knowledge-Based Artificial Neural Network* (KBANN) [172], Fu's *Knowledge-Based Connectionist Neural Network* (KBCNN) [42], Cloete's $VL_1 ANN$ [28], and the *Fuzzy Neural Network* (FuNN) of Kasabov *et al.* [81]. The rule-base is sometimes constrained, so that its implementation in the network form is straightforward [46].

Knowledge-Based Design of Recurrent ANNs

Training algorithms for RNNs are less efficient than those used for training feedforward networks [17]. Thus, the use of prior knowledge–that can improve both training and generalization performance–becomes quite important.

The most common KBD technique for RNNs is based on representing prior knowledge in the form of a deterministic finite-state automaton (DFA) [2, 125, 127]. The DFA can also be designed using learning from examples [94]. The next step is to transform the DFA into an RNN: the state-variables are realized as neurons, and the state-transitions as suitable connections between the neurons. This yields what is known as the *orthogonal internal representation* [124, 127], i.e., at every time step, only one neuron has a value ≈ 1, while all the others have values ≈ 0. Variants of this technique include: applying it to KBD of radial basis function RNNs [41]; and using gradient information in the weight space in the direction of the prior knowledge [163].

A fundamental drawback of these approaches stems from the fact that RNNs are continuous-valued and are therefore inherently different from discrete-valued mechanisms such as DFAs [86, 87]. Furthermore, the solution of certain problems, such as recognizing non-regular languages (see Section 5.3.1), cannot be represented in the form of a standard DFA at all.

Fuzzy finite-state automata (FFA) can be regarded as a continuous-valued generalization of DFAs, as state variables take values in the continuum $[0, 1]$, rather than in $\{0, 1\}$. A natural extension of the DFA-to-RNN KBD technique is based on representing the prior knowledge as an FFA, and transforming this into an RNN. This is carried out either using an intermediate FFA-to-DFA conversion (and then applying the DFA-to-RNN method [128]), or using a direct FFA-to-RNN transformation [51]. However, FFAs may include ambiguities that make the RNN implementation difficult [51].

We note in passing that RNNs have also been analyzed using the theory of dynamic systems [15, 17, 139, 140, 181], and that this approach may assist in developing new KBD methods [139].

Most of the existing neuro-fuzzy models are not relevant to KBN in RNNs, as feedback connections are rarely used [83].

1.5 The FARB: A Neuro-fuzzy Equivalence

In this work, we introduce a novel FRB, referred to as the *Fuzzy All-permutations Rule-Base* (FARB). We show that inferring the FARB, using standard tools from fuzzy logic theory,[5] yields an input-output relationship that is *mathematically equivalent* to that of a standard ANN. Conversely, every standard ANN has an equivalent FARB. We provide an *explicit* transformation T such that

$$T(\text{ANN}) = \text{FARB} \quad \text{and} \quad T^{-1}(\text{FARB}) = \text{ANN}. \tag{1.6}$$

Eq. (1.6) implies a bidirectional flow of information between an ANN and a corresponding FARB. This has several advantages:

- It enables KE from standard ANNs, in the form of standard fuzzy rules. Given an ANN, (1.6) immediately yields a suitable FARB with the same IO mapping, thus providing a *symbolic* representation of the ANN functioning.
- It enables KBD of ANNs. If the prior knowledge is stated as a FARB, the corresponding ANN follows immediately from (1.6). Note that the resulting ANN has exactly the same IO mapping as that of the original FARB.
- It enables the application of methods and tools from the field of ANNs to FARBs, and vice versa. Indeed, since the equivalence is valid for standard FRBs and standard ANNs,[6] any method used in one domain can be used in the other domain.
- It is applicable to a large range of ANN architectures, including feedforward and recurrent nets, regardless of the specific parameter values, connectivity, and network size.

---

[5] *e.g.*, Gaussian membership functions, center of gravity defuzzification.
[6] Which is not the case for most neuro-fuzzy models.

12   Introduction

Bidirectional knowledge transformation between ANNs and FRBs has many potential applications. For example, there exist many approaches for ANN pruning and for rule-base simplification. Equivalence (1.6) enables the use of ANN pruning techniques on FARBs, and the application of FARB simplification methods on ANNs. To demonstrate this, consider an algorithm that simplifies FARBs. This immediately yields an algorithm for simplifying ANNs as follows: (1) given an ANN, Compute FARB $= T(\text{ANN})$; (2) use the algorithm to simplify the FARB to, say, FARB'; and (3) compute ANN' $= T^{-1}(\text{FARB'})$. Then, ANN' is a simplified version of ANN.

Another potential application is *knowledge refinement* using the FARB–ANN equivalence. This can be done as follows: (1) state the initial knowledge as a FARB;[7] (2) obtain the corresponding ANN by computing ANN $= T^{-1}(\text{FARB})$; (3) train the network using the given data to obtain a modified network, say, ANN'; and (4) use the FARB–ANN equivalence again to extract the refined knowledge base, by calculating FARB' $= T(\text{ANN'})$.

The remainder of this book is organized as follows. In Chapter 2, we formally define the FARB and provide a closed-form formula for its input-output (IO) mapping. In Chapter 3, we show that the IO mapping of various ANNs is mathematically equivalent to the IO mapping of the FARB. We demonstrate using simple examples how this can be used to develop a new approach to KBN in ANNs. The remainder of the book is devoted to scaling this approach to larger scale ANNs. Given an ANN, the transformation $T(\text{ANN}) = \text{FARB}$ immediately yields a symbolic representation of the ANN input-output mapping. The comprehensibility of this representation can be increased by simplifying the FARB. A simplification procedure is presented in Chapter 4. In Chapter 5, the FARB is used for KE from ANNs that were trained to solve several benchmark problems. Chapter 6 describes two new KBD methods, and demonstrates their usefulness by designing ANNs that solve language recognition problems. In particular, we use our approach to systematically design an RNN that solves the AB language recognition problem. It is important to note that standard KBD methods of RNNs, that are based on converting a DFA that solves the problem into an RNN, cannot be used here. Indeed, since the AB language is context-free, there does not exist a DFA that solves the associated recognition problem. Chapter 7 concludes and discusses several potential directions for further research.

---

[7] As noted above, FRBs are particularly suitable for representing partial and self-contradicting knowledge due to their inherent ability to handle vagueness and uncertainty [122].

2 The FARB

In this chapter, we formally define the main tool developed in this work: the *fuzzy all-permutations rule-base* (FARB). We show that the special structure of the FARB implies that its IO mapping can be described using a relatively simple closed-form formula.

To motivate the definition of the FARB, we first consider a simple example adapted from [106; 107].

Example 2.1. Consider the following four-rule FRB:

R_1: If x_1 is *smaller than* 5 and x_2 *equals* 1, Then $f = -4$,
R_2: If x_1 is *larger than* 5 and x_2 *equals* 1, Then $f = 0$,
R_3: If x_1 is *smaller than* 5 and x_2 *equals* 7, Then $f = 2$,
R_4: If x_1 is *larger than* 5 and x_2 *equals* 7, Then $f = 6$.

Assume that the pair of terms {*equals* 1, *equals* 7} in the rules are modeled using the Gaussian membership function (MF):

$$\mu_{=k}(y) := \exp\left(-\frac{(y-k)^2}{2\sigma^2}\right), \tag{2.1}$$

with $k = 1$ and $k = 7$, and that the terms {*larger than* k, *smaller than* k} are modeled using the Logistic functions:

$$\mu_{>k}(y) := \frac{1}{1 + \exp\left(-\alpha(y-k)\right)} \quad \text{and} \quad \mu_{<k}(y) := \frac{1}{1 + \exp\left(\alpha(y-k)\right)},$$

with $\alpha > 0$ (see Fig. 2.1).

Applying the product-inference rule, singleton fuzzifier, and the center of gravity (COG) defuzzifier [164] to this rule-base yields: $f(\mathbf{x}) = u(\mathbf{x})/d(\mathbf{x})$, where

$$\begin{aligned}
u(\mathbf{x}) &= -4\mu_{<5}(x_1)\mu_{=1}(x_2) + 2\mu_{<5}(x_1)\mu_{=7}(x_2) + 6\mu_{>5}(x_1)\mu_{=7}(x_2), \\
d(\mathbf{x}) &= \mu_{<5}(x_1)\mu_{=1}(x_2) + \mu_{>5}(x_1)\mu_{=1}(x_2) + \mu_{<5}(x_1)\mu_{=7}(x_2) \\
&\quad + \mu_{>5}(x_1)\mu_{=7}(x_2).
\end{aligned} \tag{2.2}$$

E. Kolman, M. Margaliot: Knowledge-Based Neurocomputing, STUDFUZZ 234, pp. 13–19.
springerlink.com   © Springer-Verlag Berlin Heidelberg 2009

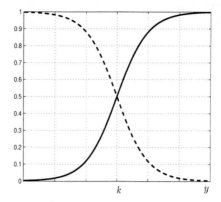

Fig. 2.1. MFs $\mu_{>k}(y)$ (solid) and $\mu_{<k}(y)$ (dashed) as a function of y

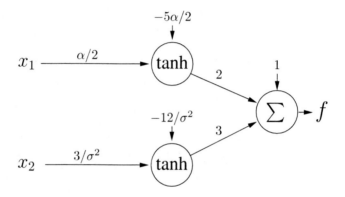

Fig. 2.2. Graphical representation of the FRB input-output mapping

Rewriting $u(\mathbf{x})$ as

$$u(\mathbf{x}) = (1 - 2 - 3)\mu_{<5}(x_1)\mu_{=1}(x_2) + (1 + 2 - 3)\mu_{>5}(x_1)\mu_{=1}(x_2)$$
$$+ (1 - 2 + 3)\mu_{<5}(x_1)\mu_{=7}(x_2) + (1 + 2 + 3)\mu_{>5}(x_1)\mu_{=7}(x_2),$$

and using (2.2) yields

$$f(\mathbf{x}) = u(\mathbf{x})/d(\mathbf{x})$$
$$= 1 + 2\frac{(\mu_{>5}(x_1) - \mu_{<5}(x_1))(\mu_{=1}(x_2) + \mu_{=7}(x_2))}{(\mu_{>5}(x_1) + \mu_{<5}(x_1))(\mu_{=1}(x_2) + \mu_{=7}(x_2))}$$
$$+ 3\frac{(\mu_{>5}(x_1) + \mu_{<5}(x_1))(-\mu_{=1}(x_2) + \mu_{=7}(x_2))}{(\mu_{>5}(x_1) + \mu_{<5}(x_1))(\mu_{=1}(x_2) + \mu_{=7}(x_2))}.$$

A direct calculation shows that for our MFs:

$$\frac{\mu_{>k}(y) - \mu_{<k}(y)}{\mu_{>k}(y) + \mu_{<k}(y)} = \tanh(\frac{\alpha(y-k)}{2}),$$

$$\frac{\mu_{=a}(y) - \mu_{=b}(y)}{\mu_{=a}(y) + \mu_{=b}(y)} = \tanh(\frac{(2y - a - b)(a - b)}{4\sigma^2}),$$

so

$$f(\mathbf{x}) = 1 + 2\tanh((x_1 - 5)\alpha/2) + 3\tanh(3(x_2 - 4)/\sigma^2).$$

Thus, the IO mapping $(x_1, x_2) \rightarrow f(x_1, x_2)$ of the FRB is *mathematically equivalent* to the IO mapping of a feedforward ANN with two hidden neurons (see Fig. 2.2). Conversely, the ANN depicted in Fig. 2.2 is mathematically equivalent to the aforementioned FRB. Note that the network parameters are directly related to the parameters of the FRB, and vice versa. □

2.1 Definition

We say that a function $g(\cdot) : \mathbb{R} \rightarrow \mathbb{R}$ is *sigmoid* if g is continuous, and the limits $\lim_{y \rightarrow -\infty} g(y)$ and $\lim_{y \rightarrow +\infty} g(y)$ exist. Example 2.1 motivates the search for an FRB whose IO mapping is equivalent to a linear combination of sigmoid functions, as this is the mapping of an ANN with a single hidden layer. This is the *FARB*. For the sake of simplicity, we consider a FARB with output $f \in \mathbb{R}$; the generalization to the case of multiple outputs is straightforward.

Definition 2.2. *An FRB with time-varying inputs $x_1(t), \ldots, x_m(t)$ and output $f(t)$ is called a FARB if the following three conditions hold.*
1. Every input variable $x_i(t)$ is characterized by two verbal terms, say, $term^i_-$ and $term^i_+$. These terms are modeled using two membership functions (MFs): $\mu^i_-(\cdot)$ and $\mu^i_+(\cdot)$. Define

$$\beta_i(y) := \frac{\mu^i_+(y) - \mu^i_-(y)}{\mu^i_+(y) + \mu^i_-(y)}.$$

The MFs satisfy the following constraint: there exist sigmoid functions $g_i(\cdot) : \mathbb{R} \rightarrow \mathbb{R}$, and $q_i, r_i, u_i, v_i \in \mathbb{R}$ such that

$$\beta_i(y) = q_i g_i(u_i y - v_i) + r_i, \qquad \text{for all } y \in \mathbb{R}. \tag{2.3}$$

2. The rule-base contains 2^m fuzzy rules spanning, in their If-part, all the possible verbal assignments of the m input variables.

3. *There exist $a_i(t) : \mathbb{R} \to \mathbb{R}$, $i = 0, 1, \ldots, m$, such that the Then-part of each rule is a combination of these functions. Specifically, the rules are:*

$$R_1 : \text{If } (x_1(t) \text{ is } term_-^1) \,\&\, (x_2(t) \text{ is } term_-^2) \,\&\ldots\&\, (x_m(t) \text{ is } term_-^m)$$
$$\text{Then } f(t) = a_0(t) - a_1(t) - a_2(t) - \cdots - a_m(t),$$

$$R_2 : \text{If } (x_1(t) \text{ is } term_+^1) \,\&\, (x_2(t) \text{ is } term_-^2) \,\&\ldots\&\, (x_m(t) \text{ is } term_-^m)$$
$$\text{Then } f(t) = a_0(t) + a_1(t) - a_2(t) - \cdots - a_m(t),$$

$$\vdots$$

$$R_{2^m} : \text{If } (x_1(t) \text{ is } term_+^1) \,\&\, (x_2(t) \text{ is } term_+^2) \,\&\ldots\&\, (x_m(t) \text{ is } term_+^m)$$
$$\text{Then } f(t) = a_0(t) + a_1(t) + a_2(t) + \cdots + a_m(t), \tag{2.4}$$

where & denotes "and". Note that the signs in the Then-part are determined in the following manner: if the term characterizing $x_i(t)$ in the If-part is $term_+^i$, then in the Then-part, $a_i(t)$ is preceded by a plus sign; otherwise, $a_i(t)$ is preceded by a minus sign.

Summarizing, the FARB is a standard FRB satisfying several additional constraints: each input variable is characterized by two verbal terms; the terms are modeled using MFs that satisfy (2.3); the rule-base contains exactly 2^m rules; and the values in the Then-part of the rules are not independent, but rather they are a linear combination of the $m + 1$ functions $a_0(t), \ldots, a_m(t)$. $\square$

As we will see below, the IO mapping of the FARB is a weighted sum of the g_is. We will be particularly interested in the case where each g_i is a function that is commonly used as an activation function in ANNs (e.g., the hyperbolic tangent function, the Logistic function).

Remark 2.3. It is easy to verify that the FRB defined in Example 2.1 is a FARB with $m = 2$, $a_0(t) \equiv 1$, $a_1(t) \equiv 2$, and $a_2(t) \equiv 3$. $\square$

Remark 2.4. It may seem that the constraint (2.3) is very restrictive. In fact, several MFs that are commonly used in FRBs satisfy (2.3). Relevant examples include the following.

1. If the terms $\{term_+, term_-\}$ are $\{equals\ k_1, equals\ k_2\}$, respectively, where the term *equals k* is modeled using the Gaussian MF (2.1), then it is easy to verify that

$$\beta(y) = \tanh(ay - b), \tag{2.5}$$

with

$$a := (k_1 - k_2)/(2\sigma^2),$$
$$b := (k_1^2 - k_2^2)/(4\sigma^2).$$

Thus, Eq. (2.3) holds with

$$g(z) = \tanh(z), \ u = a, \ v = b, \ q = 1, \text{ and } r = 0.$$

Note that (2.5) can also be written as

$$\beta(y) = 2\sigma(2ay - 2b) - 1,$$

where $\sigma(\cdot)$ is the *logistic function*:

$$\sigma(z) := (1 + \exp(-z))^{-1}.$$

Thus, Eq. (2.3) holds with

$$g(z) = \sigma(z), \ u = 2a, \ v = 2b, \ q = 2, \ \text{and } r = -1.$$

2. If the two MFs satisfy

$$\mu_-(y) = 1 - \mu_+(y), \tag{2.6}$$

(a common choice for two contradictory fuzzy terms), then $\beta(y) = 2\mu_+(y) - 1$, so (2.3) holds. Interesting special cases include:

a) If the terms $\{term_+, term_-\}$ are $\{equals\ k, not\ equals\ k\}$, respectively, modeled using the MFs:

$$\mu_{=k}(y) := \exp\left(-\frac{(y-k)^2}{2\sigma^2}\right), \quad \mu_{\neq k}(y) := 1 - \mu_{=k}(y),$$

then

$$\beta(y) = 2\exp\left(-\frac{(y-k)^2}{2\sigma^2}\right) - 1.$$

Thus, (2.3) holds with

$$g(z) = \exp(-z^2), \ u = \sqrt{\frac{1}{2\sigma^2}}, \ v = \sqrt{\frac{k^2}{2\sigma^2}}, \ q = 2, \ \text{and } r = -1.$$

b) If the terms $\{term_+, term_-\}$ are $\{larger\ than\ k, smaller\ than\ k\}$, respectively, modeled using the Logistic functions:

$$\mu_{>k}(y) := \sigma(\alpha(y-k)), \quad \mu_{<k}(y) := \sigma(-\alpha(y-k)), \tag{2.7}$$

with $\alpha > 0$,[1] then

$$\beta(y) = 2\sigma(\alpha(y-k)) - 1. \tag{2.8}$$

Thus, (2.3) holds with

$$g(z) = \sigma(z), \ u = \alpha, \ v = \alpha k, \ q = 2, \ \text{and } r = -1. \tag{2.9}$$

Note that (2.8) can also be written as

$$\beta(y) = \tanh(\alpha(y-k)/2),$$

which implies that (2.3) holds with

$$g(z) = \tanh(z), \ u = \alpha/2, \ v = \alpha k/2, \ q = 1, \ \text{and } r = 0.$$

---

[1] Note that the MFs in (2.7) satisfy (2.6).

18   The FARB

c) If the terms $\{term_+, term_-\}$ are $\{positive, negative\}$, respectively, modeled using:

$$\mu_{pos}(y) := \begin{cases} 0, & \text{if } -\infty < y < -\Delta, \\ (1 + y/\Delta)/2, & \text{if } -\Delta \le y \le \Delta, \\ 1, & \text{if } \Delta < y < \infty, \end{cases}$$

$$\mu_{neg}(y) := 1 - \mu_{pos}(y), \tag{2.10}$$

with $\Delta > 0$, then

$$\beta(y) = 2\sigma_L(y/(2\Delta) + 1/2) - 1,$$

where $\sigma_L(\cdot)$ is the *standard piecewise linear Logistic function*:

$$\sigma_L(y) := \begin{cases} 0, & \text{if } -\infty < y < 0, \\ y, & \text{if } 0 \le y \le 1, \\ 1, & \text{if } 1 < y < \infty. \end{cases} \tag{2.11}$$

This implies that (2.3) holds with

$$g(z) = \sigma_L(z), \ u = 1/(2\Delta), \ v = -1/2, \ q = 2, \text{ and } r = -1.$$

d) If the terms $\{term_+, term_-\}$ are $\{larger\ than\ k,\ smaller\ than\ k\}$, modeled using:

$$\mu_{>k}(y) := \mu_{pos}(y - k), \text{ and } \mu_{<k}(y) := \mu_{neg}(y - k), \tag{2.12}$$

with μ_{pos}, μ_{neg} defined in (2.10), then

$$\beta(y) = 2\sigma_L((y - k)/(2\Delta) + 1/2) - 1, \tag{2.13}$$

so (2.3) holds with

$$g(z) = \sigma_L(z), \ u = 1/(2\Delta), \ v = \frac{k - \Delta}{2\Delta}, \ q = 2, \text{ and } r = -1.$$

Summarizing, all the above verbal terms can be modeled using MFs that satisfy (2.3). $\qquad\square$

2.2 Input-Output Mapping

The next result provides a closed-form formula for the IO mapping of the FARB.

Theorem 2.5. *Applying the product-inference rule, singleton fuzzifier, and the COG defuzzifier to a FARB yields the output:*

$$f = a_0(t) + \sum_{i=1}^{m} r_i a_i(t) + \sum_{i=1}^{m} q_i a_i(t) g_i(u_i x_i(t) - v_i). \tag{2.14}$$

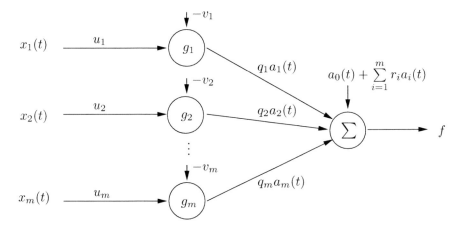

Fig. 2.3. Graphical representation of the FARB input-output mapping

PROOF. For the sake of notational convenience, we omit from hereon the dependence of the variables on t. Definition 2.2 implies that inferring the FARB yields $f(\mathbf{x}) = u(\mathbf{x})/d(\mathbf{x})$, with

$$u(\mathbf{x}) := (a_0 + a_1 + a_2 + \ldots + a_m)\mu_+^1(x_1)\mu_+^2(x_2)\ldots\mu_+^m(x_m)$$
$$+ (a_0 - a_1 + a_2 + \ldots + a_m)\mu_-^1(x_1)\mu_+^2(x_2)\ldots\mu_+^m(x_m)$$
$$\vdots$$
$$+ (a_0 - a_1 - a_2 - \ldots - a_m)\mu_-^1(x_1)\mu_-^2(x_2)\ldots\mu_-^m(x_m),$$

and

$$d(\mathbf{x}) := \mu_+^1(x_1)\mu_+^2(x_2)\ldots\mu_+^m(x_m) + \mu_-^1(x_1)\mu_+^2(x_2)\ldots\mu_+^m(x_m)$$
$$+ \ldots + \mu_-^1(x_1)\mu_-^2(x_2)\ldots\mu_-^m(x_m),$$

where both u and d include 2^m terms.

Let

$$p(\mathbf{x}) := a_0 + \sum_{i=1}^{m} a_i \beta_i(x_i). \qquad (2.15)$$

It is easy to verify, by expanding the sum, that $p = u/d$. Thus, $f = p$. Eqs (2.3) and (2.15) yield $p(\mathbf{x}) = a_0 + \sum_{i=1}^{m} a_i \left(q_i g_i(u_i x_i - v_i) + r_i\right)$, and this completes the proof. □

Eq. (2.14) implies that the FARB output f can be obtained by first feeding the (scaled and biased) inputs $u_i x_i(t) - v_i$ to a layer of units computing the activation functions $g_i(\cdot)$, and then computing a weighted (and biased) sum of the units outputs (see Fig. 2.3). Applications of this resemblance between FARBs and ANNs for knowledge-based neurocomputing are presented in the following chapter.

3 The FARB–ANN Equivalence

In this chapter, we consider several special cases of the FARB. In each of these cases, the IO mapping of the FARB is mathematically equivalent to that of a specific type of ANN. This provides a *symbolic* representation of the ANN functioning. As a corollary, we obtain results that can be used for KE from, and KBD of, ANNs. Simple examples are used to demonstrate the main ideas. The remainder of this work is devoted to extending this approach to larger-scale networks.

3.1 The FARB and Feedforward ANNs

Consider a feedforward ANN with: inputs $z_1, \ldots, z_k$, n hidden neurons with activation function $h(\cdot)$, weights w_{ij} and a single output o (see Fig. 3.1). For notational convenience, denote $w_{i0} = b_i$, and $y_i := \sum_{j=1}^{k} w_{ij} z_j$. Then

$$o = c_0 + \sum_{i=1}^{n} c_i h \left(y_i + w_{i0}\right). \tag{3.1}$$

Consider (2.14) in the particular case where x_i and a_i are time-invariant,[1] so,

$$f = a_0 + \sum_{i=1}^{m} a_i r_i + \sum_{i=1}^{m} a_i q_i g_i (u_i x_i - v_i). \tag{3.2}$$

Comparing (3.1) and (3.2) yields the following results. We use the notation $[j:k]$ for the set $\{j, j+1, \ldots, k\}$.

Corollary 3.1. (*KE from Feedforward ANNs*)
Consider the feedforward ANN (3.1). Let f denote the output of a FARB with: MFs that satisfy (2.3) with $g_i = h$, $m = n$ inputs $x_i = y_i/u_i$, and parameters

---

[1] Note that in this special case, the FARB reduces to the APFRB defined in [90; 91].

E. Kolman, M. Margaliot: Knowledge-Based Neurocomputing, STUDFUZZ 234, pp. 21–35.
springerlink.com © Springer-Verlag Berlin Heidelberg 2009

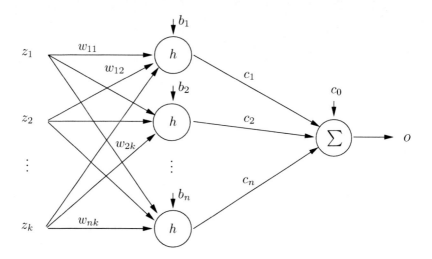

Fig. 3.1. Feedforward ANN with a single hidden layer

$$v_i = -w_{i0}, \quad a_i = c_i/q_i \quad \text{for } i \in [1{:}n],$$
$$a_0 = c_0 - \sum_{i=1}^{n} c_i r_i/q_i.$$

Then $f = o$.

Corollary 3.1 implies that given a feedforward ANN in the form (3.1), we can immediately design a FARB whose IO mapping is *mathematically equivalent* to that of the ANN. This provides a *symbolic* representation of the ANN's IO mapping. The next example demonstrates this.

3.1.1 Example 1: Knowledge Extraction from a Feedforward ANN

Consider the two-input-one-output ANN depicted in Fig. 3.2. The ANN output is given by

$$o = \sigma(4z_1 + 4z_2 - 2) - \sigma(4z_1 + 4z_2 - 6). \tag{3.3}$$

We assume that the inputs are binary, that is, $z_i \in \{0, 1\}$, and declare the ANN decision to be one if $o > 1/2$, and zero otherwise. Unfortunately, both Fig. 3.2 and Eq. (3.3) provide little insight as to what the ANN is actually computing. In this simple case, however, we can easily calculate f for the four possible input combinations, and find that $o > 1/2$ if and only if (iff) $(z_1 \text{ xor } z_2) = 1$, so the ANN is computing the xor function.

We now transform this ANN into an equivalent FARB. Denote $y_i := 4z_1 + 4z_2$, $i = 1, 2$, that is, the inputs to the hidden neurons, and rewrite (3.3) as

$$o = \sigma(y_1 - 2) - \sigma(y_2 - 6).$$

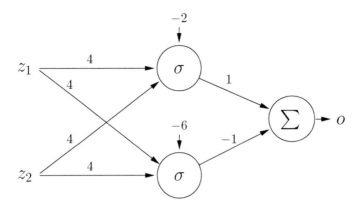

Fig. 3.2. A feedforward ANN

Note that this is in the form (3.1) with

$$n = 2, \ c_0 = 0, \ c_1 = 1, \ c_2 = -1, \ w_{10} = -2, \ w_{20} = -6, \text{ and } h = \sigma.$$

We apply Corollary 3.1 to design a FARB with $m = n = 2$ inputs x_i, and IO mapping $(x_1, x_2) \to f(x_1, x_2)$ that is equivalent to the mapping $(y_1, y_2) \to o(y_1, y_2)$. Suppose that for input x_i, $i = 1, 2$, we choose to use the verbal terms *larger than* k_i and *smaller than* k_i, modeled using (2.7) with, say, $\alpha = 4$. Then (2.8) yields

$$\beta(y_i) = 2\sigma(4(y_i - k_i)) - 1,$$

so the parameters in (2.3) are

$$q_i = 2, \ g_i = \sigma, \ u_i = 4, \ v_i = 4k_i, \ r_i = -1. \tag{3.4}$$

Applying Corollary 3.1 implies that the equivalent FARB has inputs $x_1 = y/4$, $x_2 = y/4$, and parameters: $a_0 = 0$, $a_1 = 1/2$, $a_2 = -1/2$, $v_1 = 2$, and $v_2 = 6$, so (3.4) yields $k_1 = 1/2$ and $k_2 = 3/2$. Summarizing, the equivalent FARB is:

R_1: If $y/4$ is *smaller than* $1/2$ & $y/4$ is *smaller than* $3/2$, Then $f = 0$,
R_2: If $y/4$ is *smaller than* $1/2$ & $y/4$ is *larger than* $3/2$, Then $f = -1$,
R_3: If $y/4$ is *larger than* $1/2$ & $y/4$ is *smaller than* $3/2$, Then $f = 1$,
R_4: If $y/4$ is *larger than* $1/2$ & $y/4$ is *larger than* $3/2$, Then $f = 0$,

where '&' denotes 'and'.

This FARB provides a *symbolic* description of the ANN's IO mapping. It can be further simplified as follows. Rule R_2 is self-contradicting and can be deleted. The remaining three rules can be summarized as:

If $z_1 + z_2$ is *larger than* $1/2$ and *smaller than* $3/2$, Then $f = 1$;
Else $f = 0$.

Recalling that $z_i \in \{0, 1\}$, we see that this single rule is indeed an intuitive description of the function $f(z_1, z_2) = z_1$ xor z_2. Thus, the transformation from

24   The FARB–ANN Equivalence

an ANN to an equivalent FARB yields a comprehensible representation of the network operation.   □

The next result is the converse of Corollary 3.1, namely, it states that given a FARB, we can represent its IO mapping in the form of an ANN.

Corollary 3.2. *(**KBD of Feedforward ANNs**)*
Consider a FARB with m inputs $x_1, \ldots, x_m$ and output f. Suppose that (2.3) holds for all $i \in [1, m]$ such that $g_1 = \cdots = g_m$. Define $n = m$, $y_i = u_i x_i$, $w_{i0} = -v_i$, $c_i = a_i q_i$, for $i \in [1 : n]$, $c_0 = a_0 + \sum_{i=1}^{n} a_i r_i$, and the activation function $h = g_1$. Then the FARB's output f satisfies $f = o$, where o is given by (3.1).

This result provides a useful tool for KBD of feedforward ANNs. The next example demonstrates this.

3.1.2   Example 2: Knowledge-Based Design of a Feedforward ANN

Consider the problem of designing an ANN with two binary inputs z_1, $z_2 \in \{0, 1\}$, and a single output $f = \text{not}(z_1 \text{ xor } z_2)$. In other words, the ANN should compute the xornot function.

Suppose that our initial knowledge is the truth table of the function xornot (z_1, z_2), shown graphically in Fig. 3.3. It is easy to see that the two input combinations for which $\text{xornot}(z_1, z_2) = 1$ (denoted by $\times$) are inside the region bounded by the two parallel lines $z_2 - z_1 = 1/2$ and $z_2 - z_1 = -1/2$. Hence, letting $p := z_2 - z_1$, we can state the required operation in symbolic form as:

$$\text{If } -1/2 < p < 1/2 \text{ then } f = 1; \text{ otherwise } f = 0.$$

Motivated by (2.4), we restate this using the following set of rules:

R_1: If p is *smaller than* $1/2$ and p is *larger than* $-1/2$, Then $f = 1$,
R_2: If p is *smaller than* $1/2$ and p is *smaller than* $-1/2$, Then $f = 0$,
R_3 : If p is *larger than* $1/2$ and p is *larger than* $-1/2$, Then $f = 0$.

To transform this FRB into a FARB, we must first find a_i, $i = 0, 1, 2$, such that:

$$a_0 - a_1 + a_2 = 1, \quad a_0 - a_1 - a_2 = 0, \quad \text{and} \quad a_0 + a_1 + a_2 = 0.$$

This yields

$$a_0 = 0, \; a_1 = -1/2, \text{ and } a_2 = 1/2. \tag{3.5}$$

We also need to add the fourth rule:

R_4: If p is *larger than* $1/2$ and p is *smaller than* $-1/2$,
    Then $f = a_0 + a_1 - a_2 = -1$.

Note that the degree-of-firing of this rule will always be very low, suggesting that adding it to the rule-base is harmless.

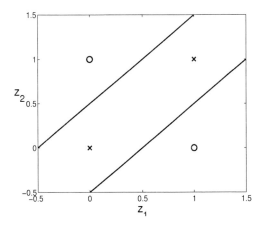

Fig. 3.3. The function xornot(z_1, z_2). ∘ denotes zero and × denotes one. Also shown are the lines $z_2 - z_1 = 1/2$, and $z_2 - z_1 = -1/2$.

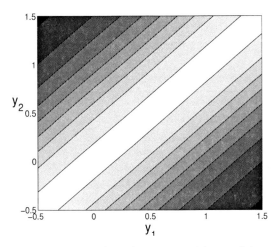

Fig. 3.4. Contour plot of $o = (\tanh(y_2 - y_1 + 1/2) - \tanh(y_2 - y_1 - 1/2))/2$

Suppose that we model the linguistic terms {*larger than k, smaller than k*} as in (2.7) with, say, $\alpha = 2$. Then (2.8) yields $\beta(y) = \tanh(y-k)$, so (2.3) holds with:

$$q = 1, \ g = \tanh, \ u = 1, \ v = k, \text{ and } r = 0.$$

Our four-rule FRB is now a FARB with $m = 2$ inputs $x_1 = x_2 = p$, $g_i = \tanh$, $u_i = 1$, $v_1 = 1/2$, $v_2 = -1/2$, $r_i = 0$, $q_i = 1$, and the a_is given in (3.5).

Applying Corollary 3.2 shows that the IO mapping $(x_1, x_2) \to f(x_1, x_2)$ of this FARB is equivalent to the mapping $(y_1, y_2) \to o(y_1, y_2)$ of the ANN:

$$o = -\frac{1}{2}\tanh(y_1 - 1/2) + \frac{1}{2}\tanh(y_2 + 1/2), \tag{3.6}$$

where $y_1 = y_2 = p = z_2 - z_1$. It is clear that (3.6) describes an ANN with two neurons in the hidden layer, and a hyperbolic tangent as the activation function.

To make the output binary, we declare the ANN decision to be one if $o > 1/2$ and zero, otherwise. A contour plot of o (see Fig. 3.4) shows that o can indeed be used to compute the xornot function.

Summarizing, we were able to systematically design a suitable ANN by stating our initial knowledge as a FARB, and then using the mathematical equivalence between the FARB and a feedforward ANN. □

The FARB–ANN equivalence can also be used for KBN in RNNs.

3.2 The FARB and First-Order RNNs

Consider a first-order RNN [127] with hidden neurons $s_1, \ldots, s_k$, activation function h, input neurons $s_{k+1}, \ldots, s_n$, and weights w_{ij} (see Fig. 3.5). Denoting, for convenience, $y_i(t) := \sum_{j=1}^{n} w_{ij} s_j(t)$, $s_0(t) \equiv 1$, and $w_{i0} = b_i$ yields

$$s_i(t+1) = h\left(y_i(t) + w_{i0}\right)$$
$$= h(\sum_{j=0}^{n} w_{ij} s_j(t)), \tag{3.7}$$

for all $i \in [1{:}k]$. We now consider several types of FARBs with an equivalent IO mapping.

3.2.1 First Approach

Consider a two-rule FARB (that is, $m = 1$) with time-invariant parameters a_0, a_1, q_1, and r_1 satisfying:

$$a_1 q_1 = 1, \quad \text{and} \quad a_0 + r_1 a_1 = 0. \tag{3.8}$$

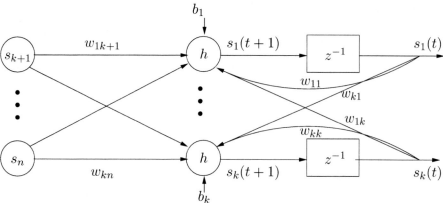

Fig. 3.5. A first-order RNN

The FARB and First-Order RNNs   27

Substituting (3.8) in (2.14) shows that the output of this FARB is

$$f = g_1(u_1 x_1(t) - v_1).\tag{3.9}$$

Comparing (3.9) and (3.7) yields the following results.

Corollary 3.3. (*KE from a First-Order RNN*)
Consider the first-order RNN (3.7). Let f denote the output of a two-rule FARB with: MFs that satisfy (2.3) with $g_1 = h$, input $x_1(t) = y_i(t)/u_1$, parameters that satisfy (3.8), and $v_1 = -w_{i0}$. Then

$$f = s_i(t+1).$$

This result can be used for KE from RNNs. The next example demonstrates this.

3.2.2   Example 3: Knowledge Extraction from a Simple RNN

Consider the RNN
$$s_i(t+1) = \sigma(w_{i1} s_1(t) + w_{i0}).\tag{3.10}$$

Note that this is in the form (3.7) with $n = 1$, $h = \sigma$, and $y_1 = w_{i1} s_1(t)$.

Corollary 3.3 can be applied to yield a single-input two-rule FARB with an equivalent IO mapping. Suppose that we use the fuzzy terms {*larger than k, smaller than k*}, modeled using the MFs defined in (2.7), with $\alpha > 0$. Then (2.9) implies that

$$g(z) = \sigma(z), \ u_1 = \alpha, \ v_1 = \alpha k, \ q_1 = 2, \text{ and } r_1 = -1,$$

so (3.8) yields
$$a_0 = a_1 = 1/2.$$

Applying Corollary 3.3 implies that the equivalent FARB is:

R_1: If $\frac{w_{i1} s_1(t)}{\alpha}$ is *larger than* $\frac{-w_{i0}}{\alpha}$, Then $s_i(t+1) = 1$,
R_2: If $\frac{w_{i1} s_1(t)}{\alpha}$ is *smaller than* $\frac{-w_{i0}}{\alpha}$, Then $s_i(t+1) = 0$.

In other words, the IO mapping of this FARB is identical to the mapping given in (3.10). This provides a *symbolic* representation of the RNN (3.10).   □

The next result is the converse of Corollary 3.3.

Corollary 3.4. (*KBD of a First-Order RNN*)
Consider a two-rule FARB with input $x_1(t)$, output $f(t)$, and parameters satisfying (3.8). Define $y_i(t) = u_1 x_1(t)$, $w_{i0} = -v_1$, and the activation function $h = g_1$. Then the FARB's output satisfies $f = s_i(t+1)$, where $s_i(t+1)$ is given by (3.7).

Corollaries 3.3 and 3.4 show that the IO mapping of every neuron in a first-order RNN is equivalent to that of a FARB with two rules. However, a two-rule FARB is usually too simple to provide useful information on the equivalent RNN. The next section describes an alternative approach.

28 The FARB–ANN Equivalence

3.2.3 Second Approach

Let $h^{-1}(\cdot)$ denote the inverse of $h(\cdot)$. We assume that the inverse exists either globally, or at least in some relevant operation domain of the network. We re-state (3.7) as follows. For any $i \in [1:k]$:

$$h^{-1}(s_i(t+1)) = \sum_{j=0}^{n} w_{ij} s_j(t) \tag{3.11}$$

$$= w_{i0} s_0(t) + \sum_{j=1}^{k} w_{ij} s_j(t) + \sum_{j=k+1}^{n} w_{ij} s_j(t)$$

$$= w_{i0} s_0(t) + \sum_{j=1}^{k} w_{ij} h(\sum_{p=0}^{n} w_{jp} s_p(t-1)) + \sum_{j=k+1}^{n} w_{ij} s_j(t).$$

It will be useful to express this in the form:

$$h^{-1}(s_i(t+1)) = w_{i0} h(w_{00} + h^{-1}(s_0(t)))$$

$$+ \sum_{j=1}^{k} w_{ij} h(w_{j0} + \sum_{p=1}^{n} w_{jp} s_p(t-1))$$

$$+ \sum_{j=k+1}^{n} w_{ij} h(w_{j0} + h^{-1}(s_j(t))),$$

where we used the fact that $w_{j0} = 0$ for $j = 0$ and for $j = [k+1:n]$. Letting

$$\tilde{y}_j(t) := \begin{cases} \sum\limits_{p=1}^{n} w_{jp} s_p(t-1), & \text{if } j \in [1:k], \\ h^{-1}(s_i(t)), & \text{if } j = 0 \text{ or } j \in [k+1:n], \end{cases} \tag{3.12}$$

yields

$$h^{-1}(s_i(t+1)) = \sum_{j=0}^{n} w_{ij} h(\tilde{y}_j(t) + w_{j0}), \quad \text{for all } i \in [1:k]. \tag{3.13}$$

Eq. (3.13) implies that $h^{-1}(s_i(t+1))$ can be represented as the output of a suitable *feedforward* ANN (see Fig. 3.6).

Comparing (3.13) and (2.14) yields the following results.

Corollary 3.5. (*KE from a First-Order RNN*)
Consider the first-order RNN given by (3.12) and (3.13). Let $f(t)$ denote the output of a FARB with: MFs that satisfy (2.3) with $g_i = h$, $m = n$ inputs $x_j(t) = \tilde{y}_j(t)/u_j$, for $j \in [1:n]$, and parameters

$$a_0 = w_{i0} - \sum_{j=1}^{m} w_{ij} r_j/q_j, \ a_j = w_{ij}/q_j, \ and \ v_j = -w_{j0}, \quad j \in [1:n].$$

Then $f(t) = h^{-1}(s_i(t+1))$.

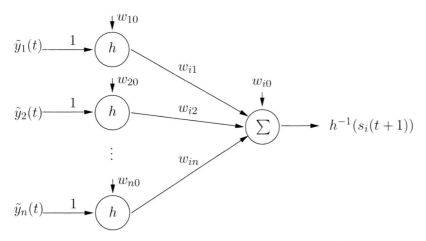

Fig. 3.6. Graphical representation of Eq. (3.13)

Corollary 3.6. (KBD of a First-Order RNN)
Consider a FARB with inputs $x_1(t), \ldots, x_m(t)$ and output $f(t)$. Define $n = m$, $\tilde{y}_j(t) = x_j(t)u_j$, and $h = g_j$, for $j \in [1:n]$, and

$$w_{i0} = a_0 + \sum_{j=1}^{m} a_j r_j, \quad w_{ij} = a_j q_j, \quad \text{and} \quad w_{j0} = -v_j, \quad j \in [1:n].$$

Then $f(t) = h^{-1}(s_i(t+1))$, where $h^{-1}(s_i(t+1))$ is given in (3.13).

Another approach for converting the RNN into a FARB or vice-versa is possible when the function g in (2.3) is piecewise-linear.

3.2.4 Third Approach

Assume that each g_i in (2.3) is a linear function (the results below can be generalized to the case where g_i is piecewise-linear), that is, there exist $\tilde{u}_i, \tilde{v}_i \in \mathbb{R}$ such that

$$g_i(u_i x_i(t) - v_i) = \tilde{u}_i x_i(t) - \tilde{v}_i. \tag{3.14}$$

Then, the FARB output given in (3.2) can be restated as

$$f(t) = a_0 + \sum_{i=1}^{m} a_i r_i + \sum_{i=1}^{m} a_i q_i (\tilde{u}_i x_i(t) - \tilde{v}_i)$$

$$= a_0 + \sum_{i=1}^{m} a_i (r_i - q_i \tilde{v}_i) + \sum_{i=1}^{m} a_i q_i \tilde{u}_i x_i(t). \tag{3.15}$$

On the other-hand, recall that the first-order RNN (3.7) can be described by (3.11), that is,

$$h^{-1}(s_i(t+1)) = \sum_{j=0}^{n} w_{ij} s_j(t). \tag{3.16}$$

Comparing (3.15) to (3.16) yields the following result.

30   The FARB–ANN Equivalence

Corollary 3.7. (KE from a First-Order RNN)
Consider the first-order RNN given by (3.16). Let $f(t)$ denote the output of a FARB with: MFs that satisfy (2.3) with g_i satisfying (3.14), $m = n$ inputs $x_j(t) = s_j(t)/\tilde{u}_j$, for $j \in [1\!:\!n]$, and parameters

$$a_0 = w_{i0} - \sum_{j=1}^{m} w_{ij}(r_j - q_j\tilde{v}_j)/q_j, \quad and \quad a_j = w_{ij}/q_j, \quad j \in [1\!:\!n].$$

Then $f(t) = h^{-1}(s_i(t+1))$.

This result provides a useful mechanism for KE from RNNs. The next example demonstrates this.

3.2.5   Example 4: Knowledge Extraction from an RNN

Consider the one-input-one-output RNN depicted in Fig. 3.7. Assume that the initial condition is $s(1) = 1$. The RNN is then described by:

$$s(t+1) = \sigma_L(s(t) + I(t) - 1), \qquad s(1) = 1. \tag{3.17}$$

We consider the case where the input is binary: $I(t) \in \{0,1\}$ for all t.

We now transform this RNN into an equivalent FARB. It will be convenient to rewrite (3.17) in the form

$$s_1(t+1) = \sigma_L(s_1(t) + s_2(t) - 1), \tag{3.18}$$

with $s_1(t) := s(t)$, and $s_2(t) := I(t)$. Note that this is a special case of (3.16) with: $k = 1$, $n = 2$, $h = \sigma_L$, and the weights:

$$w_{10} = -1, \; w_{11} = 1, \; w_{12} = 1. \tag{3.19}$$

Suppose that we characterize all the variables in the equivalent FARB using the fuzzy terms {*equals* 1, *equals* 0}, modeled using the MFs:

$$\mu_{=1}(u) := \sigma_L(u) \quad and \quad \mu_{=0}(u) := \sigma_L(1 - u). \tag{3.20}$$

The definition of $\beta(\cdot)$ yields

$$\beta(z) = \frac{\sigma_L(z) - \sigma_L(1-z)}{\sigma_L(z) + \sigma_L(1-z)}$$
$$= \sigma_L(z) - \sigma_L(1-z).$$

Note that $\sigma_L(z) = z$ for any $z \in [0,1]$. Thus, in this linear range:

$$\beta(z) = 2z - 1$$
$$= 2\sigma_L(z) - 1,$$

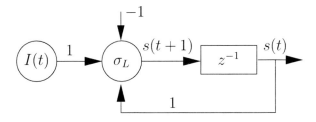

Fig. 3.7. A simple RNN

so (2.3) and (3.14) hold with:

$$g = \sigma_L, \; u = 1, \; v = 0, \; q = 2, \; r = -1, \; \tilde{u} = u = 1, \text{ and } \tilde{v} = v = 0. \quad (3.21)$$

Corollary 3.7 now implies that the RNN (3.18) is equivalent to a four-rule FARB with $m = 2$ inputs $x_1(t) = s_1(t)/\tilde{u}_1 = s(t)$, $x_2(t) = s_2(t)/\tilde{u}_2 = I(t)$, and parameters

$$a_1 = w_{11}/q_1,$$
$$a_2 = w_{12}/q_2,$$
$$a_0 = w_{10} - (w_{11}(r_1 - q_1\tilde{v}_1)/q_1 + w_{12}(r_2 - q_2\tilde{v}_2)/q_2).$$

Using (3.19) and (3.21) yields

$$a_0 = 0, \qquad a_1 = a_2 = 1/2.$$

Summarizing, the equivalent FARB is

R_1: If $s(t)$ equals 1 and $I(t)$ equals 1, Then $\sigma_L^{-1}(s(t+1)) = 1$,
R_2: If $s(t)$ equals 1 and $I(t)$ equals 0, Then $\sigma_L^{-1}(s(t+1)) = 0$,
R_3: If $s(t)$ equals 0 and $I(t)$ equals 1, Then $\sigma_L^{-1}(s(t+1)) = 0$,
R_4: If $s(t)$ equals 0 and $I(t)$ equals 0, Then $\sigma_L^{-1}(s(t+1)) = -1$.

This provides a symbolic representation of the RNN (3.17). This FARB can be further simplified as follows. Since $\sigma_L(-1) = \sigma_L(0) = 0$, and since rule R_1 is the only rule where the Then-part satisfies $\sigma_L^{-1}(s(t+1)) > 0$, the FARB can be summarized as:

If $s(t)$ equals 1 and $I(t)$ equals 1, Then $s(t+1) = \sigma_L(1) = 1$;
Else $s(t+1) = 0$.

This can be stated as:

If $I(t)$ equals 1, Then $s(t+1) = s(t)$;
Else $s(t+1) = 0$.

Using this description, it is straightforward to understand the RNN functioning. Recall that $s(t)$ is initialized to 1. It follows that $s(t)$ will remain 1 until the

32 The FARB–ANN Equivalence

first time the input is $I(t) = 0$. Once this happens, $s(t + 1)$ is set to zero, and will remain 0 from thereon. In other words, the RNN output is:

$$s(t+1) = \begin{cases} 1, & \text{if } I(\tau) = 1 \text{ for all } \tau \leq t, \\ 0, & \text{otherwise.} \end{cases}$$

If we regard the RNN as a formal language recognizer (see Section 5.3.1 below), with $s(t+1) = 1$ interpreted as Accept, and $s(t+1) = 0$ as Reject, then the RNN accepts a binary string iff it does not include any zero bits. Summarizing, the transformation into a FARB provides a comprehensible explanation of the RNN functioning. □

By comparing (3.15) to (3.16), it is also possible to derive the following result which is the converse of Corollary 3.7.

Corollary 3.8. (KBD of a First-Order RNN)
Consider a FARB with inputs $x_1(t), \ldots, x_m(t)$, output $f(t)$ and functions g_i satisfying (3.14). Define $n = m$, $s_j(t) = x_j(t)u_j$ for $j \in [1:n]$, and

$$w_{i0} = a_0 + \sum_{j=1}^{m} a_j(r_j - q_j \tilde{v}_j), \quad \text{and} \quad w_{ij} = a_j q_j, \quad j \in [1:n].$$

Then

$$f(t) = h^{-1}\left(s_i(t+1)\right),$$

where $h^{-1}\left(s_i(t+1)\right)$ is given in (3.16).

This result can be used for KBD of RNNs. The next example demonstrates this.

3.2.6 Example 5: Knowledge-Based Design of an RNN

Consider the following problem. Design an RNN that accepts a binary string $I(1), I(2), \ldots$ as an input. If the input string contained the bit 1, the RNN's output should be $s(t+1) = 0$. Otherwise, that is, if $I(1) = I(2) = \cdots = I(t) = 0$, the RNN's output should be $s(t + 1) = 1$.

Our design is based on stating the required functioning as a FARB and then using Corollary 3.8 to obtain the equivalent RNN. The design concept is simple. We initialize $s(1) = 1$. If $I(t) = 1$, then we set $s(t + 1) = 0$; if $I(t) = 0$, then s is unchanged. Thus, once a 1 appears in the input, $s(\cdot)$ will be set to zero, and remain zero from thereon.

We can state the desired behavior of $s(t)$ in the following form:

R_1: If $I(t)$ equals 1, Then $s(t + 1) = 0$,
R_2: If $I(t)$ equals 0, Then $s(t + 1) = s(t)$.

This is a two-rule FARB with a single input $x_1(t) = I(t)$, and

$$a_0(t) = s(t)/2,$$
$$a_1(t) = -s(t)/2.$$

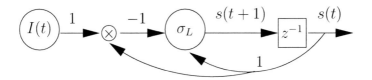

Fig. 3.8. Graphical representation of the RNN described by Eq. (3.22)

We model the fuzzy terms $\{equals\ 1, equals\ 0\}$ as in (3.20), so (2.3) and (3.14) hold with:

$$g_1 = \sigma_L,\ u_1 = 1,\ v_1 = 0,\ q_1 = 2,\ r_1 = -1,\ \tilde{u}_1 = u_1 = 1,\ \tilde{v}_1 = v_1 = 0.$$

Applying Corollary 3.8 yields $n = 1$, $s_1(t) = I(t)$, $w_{10} = s(t)$, and $w_{11} = -s(t)$. Substituting these parameters in (3.16) yields

$$\sigma_L^{-1}(s(t+1)) = s(t) - I(t)s(t),$$

or

$$s(t+1) = \sigma_L\left(s(t) - I(t)s(t)\right). \tag{3.22}$$

Clearly, this describes the dynamics of an RNN with a single neuron $s(t)$ and a single input $I(t)$ (see Fig. 3.8). Recalling that $s(1) = 1$, it is easy to verify that $s(t+1) = 0$ iff there exists $j \in [1:t]$ such that $I(j) = 1$. Otherwise, $s(t+1) = 1$. Thus, the designed RNN indeed solves the given problem. □

The next section describes a connection between the FARB and another type of RNN.

3.3 The FARB and Second-Order RNNs

Second-order RNNs, introduced by Pollack in 1987 [134], are a generalization of first-order RNNs, and, in particular, can solve problems that first-order RNNs cannot [53]. In second-order RNNs, the connection weights are linear functions of the neurons values [135]:

$$w_{ij}(t) = \sum_{l=0}^{n} w_{ijl} s_l(t).$$

Denote the hidden neurons by $s_1, \ldots, s_k$, the bias neuron by $s_0(t) \equiv 1$, and the input neurons by $s_{k+1} \ldots, s_n$, then the dynamics of a second-order RNN is given by

$$s_i(t+1) = h(\sum_{j=0}^{n} w_{ij}(t)s_j(t))$$

$$= h(\sum_{j=0}^{n}\sum_{l=0}^{n} w_{ijl} s_l(t) s_j(t)),\quad i \in [1:k], \tag{3.23}$$

34   The FARB–ANN Equivalence

where h is the activation function. The parameter w_{ij0} is the weight of the connection from neuron j to neuron i, and w_{i00} is the bias of neuron i. Note that, by definition,

$$w_{ij0} = w_{i0j}. \tag{3.24}$$

Eq. (3.23) yields

$$h^{-1}\left(s_i(t+1)\right) = \sum_{j=0}^{n}\sum_{l=0}^{n} w_{ijl}s_l(t)s_j(t)$$

$$= w_{i00} + 2\sum_{j=1}^{n} w_{ij0}s_j(t) + \sum_{j=1}^{n}\sum_{l=1}^{n} w_{ijl}s_l(t)s_j(t), \tag{3.25}$$

where the last equation follows from (3.24).

To determine a FARB with an equivalent IO mapping, consider the case where the FARB parameters satisfy:

$$a_k(t) = d_k + \sum_{j=1}^{m} d_{kj}x_j(t), \quad k \in [0{:}m]. \tag{3.26}$$

For the sake of convenience, denote $x_0(t) \equiv 1$ and $d_{k0} = d_k$, so $a_k(t) = \sum_{j=0}^{m} d_{kj}x_j(t)$. Then, (2.14) yields

$$f = \sum_{j=0}^{m} d_{0j}x_j(t) + \sum_{k=1}^{m}\sum_{j=0}^{m} d_{kj}r_k x_j(t)$$

$$+ \sum_{k=1}^{m}\sum_{j=0}^{m} d_{kj}q_k x_j(t)g_k(u_k x_k(t) - v_k).$$

Assume also that each g_k is a linear function so that (3.14) holds. Then

$$f = d_{00} + \sum_{k=1}^{m} d_{k0}(r_k - q_k\tilde{v}_k)$$

$$+ \sum_{j=1}^{m}\left(d_{0j} + d_{j0}q_j\tilde{u}_j + \sum_{k=1}^{m} d_{kj}(r_k - q_k\tilde{v}_k) \right)x_j(t)$$

$$+ \sum_{k=1}^{m}\sum_{l=1}^{m} d_{kl}q_k\tilde{u}_k x_k(t)x_l(t). \tag{3.27}$$

Comparing (3.27) and (3.25) yields the following results.

Corollary 3.9. (KE from a Second-Order RNN)
Consider the second-order RNN (3.25). Let f denote the output of a FARB with: $m = n$ inputs $x_j(t) = s_j(t)$, MFs that satisfy (2.3) and (3.14) and parameters that satisfy (3.26) and

$$d_{kl} = w_{ikl}/(q_k\tilde{u}_k),$$

$$d_{0l} + d_{l0}q_l\tilde{u}_l + \sum_{k=1}^{m}\frac{w_{ikl}}{q_k\tilde{u}_k}(r_k - q_k\tilde{v}_k) = 2w_{il0},$$

$$d_{00} + \sum_{k=1}^{m}d_{k0}(r_k - q_k\tilde{v}_k) = w_{i00},$$

for $k, l \in [1:m]$. Then $f = h^{-1}(s_i(t+1))$.

In other words, we can transform the dynamics of every neuron s_i in the second-order RNN into an equivalent FARB.

The next result is the converse of Corollary 3.9, namely, it provides a transformation from a FARB into an equivalent second-order RNN.

Corollary 3.10. (*KBD of a Second-Order RNN*)
Consider a FARB with inputs $x_1(t), \ldots, x_m(t)$, output $f(t)$, MFs such that (3.14) holds, and parameters that satisfy (3.26). Define $n = m$, $h = g_i$, $s_j(t) = x_j(t)$ for $j \in [1:n]$, and weights

$$w_{ikl} = d_{kl}q_k\tilde{u}_k,$$

$$w_{il0} = (d_{0l} + d_{l0}q_l\tilde{u}_l + \sum_{k=1}^{m}d_{kl}(r_k - q_k\tilde{v}_k))/2,$$

$$w_{i00} = d_{00} + \sum_{k=1}^{m}d_{k0}(r_k - q_k\tilde{v}_k),$$

for $k, l \in [1 : n]$. Then $f(t) = h^{-1}(s_i(t+1))$, where $h^{-1}(s_i(t+1))$ is given in (3.25).

3.4 Summary

In this chapter we studied in detail the mathematical equivalence between the FARB and various types of ANNs. We showed that the FARB–ANN equivalence holds for a large variety of ANNs, regardless of their specific architecture and parameter values. Since the FARB is a standard FRB, this enables the application of tools from the theory of ANNs to FARBs, and vice versa.

Given an ANN, we can immediately determine a suitable FARB with the same IO mapping, and thus provide a a *symbolic* description of the ANN functioning.

Conversely, consider the problem of designing an ANN for solving a given problem. In many cases, some initial knowledge about the problem domain is known. Designing a symbolic FARB based on this knowledge yields an IO mapping that can be immediately realized as a suitable ANN.

These ides were demonstrated using simple examples. Applications to larger-scale problems are studied in Chapters 5 and 6 below.

4 Rule Simplification

The interpretability of FRBs might be hampered by the existence of a large number of rules or complicated ones. Thus, any attempt to use the FARB–ANN equivalence for knowledge extraction from large-scale networks must include a systematic approach for rule reduction and simplification.

Simplification procedures usually lead to a more interpretable rule-base with the cost of a degradation in the FRB performance. This tradeoff between complexity and interpretability can be viewed as one manifestation of Zadeh's *principle of incompatibility* (see Section 1.2).

Existing procedures for simplifying rule-bases include: deleting rules with a small effect on the output determined using the singular value decomposition [10]; deleting similar, contradictory or inactive rules [77, 78, 149, 156, 191]; using the correlation between the inputs to delete or simplify the rules [153]; restricting the number of antecedents [77, 191]; and using evolution strategies to search for a simpler rule-base [78].

In the context of this work, procedures for simplifying ANNs are also quite relevant. These include: deleting connections with small weights [155]; clustering smaller sub-networks [156]; clustering the weights [156, 172] or the neurons outputs [156]. The simplification can also be integrated with the training process by constraining the architecture [47, 172], the training algorithm [47], the weight values [14, 36, 47, 156], the connectivity [14, 69], and the activation functions [36].

In this chapter, we describe a procedure for simplifying a FARB. We begin by presenting some results on the sensitivity of the FARB IO mapping to modifications of its rules, atoms, and parameters. These results will be applied below to bound the error incurred by the simplification procedure.

4.1 Sensitivity Analysis

Consider a FARB with input $\mathbf{x} \in \mathbb{R}^m$, $q := 2^m$ rules, and output $f \in \mathbb{R}$ (we omit the dependence of $\mathbf{x}$ and f on t for notational convenience). Let $t_i(\mathbf{x})$ and f_i denote the degree of firing (DOF) and the value in the Then-part, respectively, of rule i. The fuzzy inferencing process yields the output $f(\mathbf{x}) = u(\mathbf{x})/d(\mathbf{x})$, where $u(\mathbf{x}) := \sum_{i=1}^q t_i(\mathbf{x})f_i$, and $d(\mathbf{x}) := \sum_{i=1}^q t_i(\mathbf{x})$.

38 Rule Simplification

The next result analyzes the effect of perturbing the DOF of a single rule. Note that removing rule k from the FARB altogether amounts to changing $t_k(\mathbf{x})$ to $t'_k(\mathbf{x}) \equiv 0$.

Proposition 4.1. *Fix an arbitrary input* $\mathbf{x} \in \mathbb{R}^m$. *Suppose that the DOF of rule* k *is modified from* $t_k(\mathbf{x})$ *to* $t'_k(\mathbf{x})$, *and let* $f'(\mathbf{x})$ *denote the output of the modified FARB. Then,*

$$|f(\mathbf{x}) - f'(\mathbf{x})| \le c_k(\mathbf{x})|t_k(\mathbf{x}) - t'_k(\mathbf{x})|,$$

where

$$c_k(\mathbf{x}) := (1/d(\mathbf{x})) \max_{1 \le i \le q} \{|f_i(\mathbf{x}) - f_k(\mathbf{x})|\}. \tag{4.1}$$

PROOF. Denoting $\triangle t_k(\mathbf{x}) := t'_k(\mathbf{x}) - t_k(\mathbf{x})$, $u'(\mathbf{x}) := u(\mathbf{x}) + f_k \triangle t_k(\mathbf{x})$, and $d'(\mathbf{x}) := d(\mathbf{x}) + \triangle t_k(\mathbf{x})$, yields

$$\begin{aligned} f - f' &= (u/d) - (u'/d') \\ &= \frac{(u' - d'f_k)\triangle t_k}{dd'} \\ &= (f' - f_k)\frac{\triangle t_k}{d}. \end{aligned} \tag{4.2}$$

Clearly, $\min_{1 \le i \le q}\{f_i\} \le f' \le \max_{1 \le i \le q}\{f_i\}$, so (4.2) yields

$$|f - f'| \le (1/d)|\triangle t_k| \max\{|\max_{1 \le i \le q}\{f_i\} - f_k|, |\min_{1 \le i \le q}\{f_i\} - f_k|\},$$

since the DOF is always in the range $[0,1]$, $|\triangle t_k| \le 1$, and this completes the proof. $\square$

If one of the $|a_k|$s is very small, then we might expect the corresponding input x_k to have a small effect on the FARB output f. This is quantified by the following result.

Proposition 4.2. *Suppose that the atoms containing* x_k *are removed from the If-part of all the rules and that* a_k *is removed from the Then-part of all the rules. Let* $f'(\mathbf{x})$ *denote the output of the resulting FARB. Then,*

$$|f(\mathbf{x}) - f'(\mathbf{x})| \le |a_k \eta_k|, \quad \text{for all } \mathbf{x}, \tag{4.3}$$

where $\eta_k := \max_{x_k}\{r_k + q_k g_k(u_k x_k - v_k)\}$ *(see (2.3)).*

PROOF. It is easy to see that f' is also the output of a FARB, so Theorem 2.5 implies that

$$f'(\mathbf{x}) = a_0 + \sum_{\substack{i=1 \\ i \ne k}}^{m} r_i a_i + \sum_{\substack{i=1 \\ i \ne k}}^{m} q_i a_i g_i(u_i x_i - v_i).$$

Note that $f'(\mathbf{x})$ is a function of the variables $x_1, \ldots, x_{k-1}, x_{k+1}, \ldots, x_m$ only. Hence,

$$f(\mathbf{x}) - f'(\mathbf{x}) = a_k(r_k + q_k g_k(u_k x_k - v_k)),$$

which proves (4.3). $\qquad\square$

The rules in the FARB describe a partition of the input space. Roughly speaking, for any given input $\mathbf{x}$, only a few rules will fire substantially, and the other rules will have a weaker effect on $f(\mathbf{x})$. This is analyzed more rigorously in the following result.

Proposition 4.3. *Fix an arbitrary $\mathbf{x} \in \mathbb{R}^m$. Denote*

$$k(\mathbf{x}) = \arg \max_{1 \le i \le q} t_i(\mathbf{x}),$$

that is, the index of the rule with the highest DOF for the input $\mathbf{x}$, and let

$$b(\mathbf{x}) := \max_{1 \le i \le q} |f_i - f_{k(\mathbf{x})}|. \tag{4.4}$$

Then

$$|f(\mathbf{x}) - f_{k(\mathbf{x})}| \le \frac{b(\mathbf{x})}{1 + t_k(\mathbf{x}) / \sum_{i \ne k} t_i(\mathbf{x})}. \tag{4.5}$$

Note that the bound (4.5) has a clear intuitive interpretation. The difference $|f(\mathbf{x}) - f_{k(\mathbf{x})}|$ will be small if the DOF of rule k is high (with respect to the DOFs of the other rules) or if all the rules outputs are more or less equal to $f_{k(\mathbf{x})}$.

PROOF

$$\begin{aligned}
|f(\mathbf{x}) - f_{k(\mathbf{x})}| &= \left| \frac{\sum_i t_i(\mathbf{x}) f_i}{\sum_i t_i(\mathbf{x})} - f_{k(\mathbf{x})} \right| \\
&= \left| \frac{\sum_{i \ne k(\mathbf{x})} t_i(\mathbf{x})(f_i - f_{k(\mathbf{x})})}{t_{k(\mathbf{x})}(\mathbf{x}) + \sum_{i \ne k(\mathbf{x})} t_i(\mathbf{x})} \right| \\
&\le \left| \frac{\sum_{i \ne k(\mathbf{x})} t_i(\mathbf{x})}{t_{k(\mathbf{x})}(\mathbf{x}) + \sum_{i \ne k(\mathbf{x})} t_i(\mathbf{x})} \right| b(\mathbf{x}),
\end{aligned}$$

which implies (4.5). $\qquad\square$

4.2 A Procedure for Simplifying a FARB

We now describe a step-by-step procedure for simplifying a given FARB. Let $D = \{\mathbf{x}^i \in \mathbb{R}^m\}_{i=1}^s$ denote the training set.

STEP 1. For each $k \in [1:m]$, if $|a_k|$ is small (with respect to the other $|a_i|$s), remove the atoms containing x_k in the rules If-part, and remove a_k from the Then-part of all the rules. Note that Proposition 4.2 implies that the resulting error is smaller than $|a_k \eta_k|$.[1]

---

[1] In the terminology of [111], this step may be considered a form of *rule resolution*.

40   Rule Simplification

STEP 2. For rule k, $k \in [1 : q]$, compute $m_k := \max_{\mathbf{x} \in D} t_k(\mathbf{x})$, and $l_k := \max_{\mathbf{x} \in D} c_k(\mathbf{x})$ (see (4.1)). If $m_k l_k$ is small, then delete rule k from the FARB. Note that Proposition 4.1 implies that the resulting error is smaller than $m_k l_k$.

STEP 3. Let $k(\mathbf{x}) := \arg\max_i t_i(\mathbf{x})$. Compute $e := \max_{\mathbf{x} \in D} b(\mathbf{x})$ and $r := 1 + \min_{\mathbf{x} \in D} t_k(\mathbf{x}) / \sum_{i \neq k} t_i(\mathbf{x})$ (see (4.4)). If e/r is small, then set the output to $f_{k(\mathbf{x})}$ instead of $f(\mathbf{x})$. This is equivalent to replacing the COG defuzzifier with the mean of maxima (MOM) defuzzifier [78, 164]. Note that Proposition 4.3 implies that the resulting error is bounded by e/r.

STEP 4. If a specific atom (e.g., 'x_1 is smaller than 7') appears in all the q rules, then delete it from them all. It is easy to verify that in such a case, this atom has no effect on the output.

STEP 5. If the ith atom of Rule j is 'x_i is $Term^i_+$' ($Term^i_-$), and in all the other rules the ith atom is 'x_i is $Term^i_-$' ($Term^i_+$), then: (1) remove all the atoms, except for atom i, from the If-part of Rule j; (2) delete the ith atom from all the other rules; and (3) place Rule j as the first rule, and add an *Else* clause followed by all the other rules.

STEP 6. If the FARB is used for classification, define one class as the *default class*, delete all the rules whose output is this class, and add the clause: "Else, class is the default class" to the rule-base [92].

STEP 7. Recall that the jth input to the FARB is $x_j = \sum_{i=1}^{n} w_{ji} z_i$. If $|w_{ji}|$ is small, then replace the term $w_{ji} z_i$ with the term $w_{ji} \bar{z}_i$, where $\bar{z}_i$ is the expectation of z_i over the input space. This step yields a simpler description of the rule-base in terms of the original inputs z_i. It should be noted, however, that this may result in a substantial approximation error.

This completes the description of the simplification procedure. To justify STEP 5, suppose that 'x_i is $Term^i_+$' appears in Rule j and 'x_i is $Term^i_-$' appears in all the other rules. Roughly speaking, this implies that the input space is divided into two regions. In the region where $\mu^i_+(\mathbf{x})$ is high, $f(\mathbf{x}) \approx f_j$, so the other rules can be ignored. In the region where $\mu^i_+(\mathbf{x}) \approx 0$, Rule j can be ignored, and we can also ignore the identical atom i in all the other rules (see STEP 4).

In classification problems, as opposed to function approximation (or regression) problems, $|f(\mathbf{x}) - f'(\mathbf{x})|$ is not necessarily a suitable error criterion. A more appropriate measure is the number of misclassification. Thus, large changes in the output can be tolerated, as long as the actual classification remains unchanged (see Section 5.2.2 below for a modified application of STEP 2, suitable for FARBs used for classification problems).

In the next chapter, we apply the FARB–ANN equivalence for knowledge extraction from trained ANNs. This is combined with the simplification procedure described above in order to increase the comprehensibility for the extracted FARB.

5 Knowledge Extraction Using the FARB

The FARB–ANN equivalence immediately yields a representation of the IO mapping of a given trained ANN in a symbolic form. The fuzzy rules and their parameters are determined by the network structure and parameters. This transforms the knowledge embedded in the network into a set of symbolic rules stated using natural language. If necessary, the FARB can be further simplified to increase its comprehensibility.

In Chapter 3, we demonstrated the usefulness of this approach using simple examples. In this chapter, we apply this method to extract knowledge from ANNs trained to solve three benchmark problems: The Iris classification problem, the LED display recognition problem, and a language recognition problem. All these problems have been intensively discussed in the literature, and are often used as examples for demonstrating the performance of various machine learning algorithms.

5.1 The Iris Classification Problem

The Iris classification problem [39] is a well-known benchmark for machine learning algorithms. The data[1] is the description of 150 flowers as vectors in the form (z_1, z_2, z_3, z_4, v). The z_is are the size (in centimeters) of the four attributes: Petal width, Petal length, Sepal width, and Sepal length, respectively. The value v describes the flowers class (Setosa, Versicolor, or Virginica). One class is linearly separable from the other two, which are not linearly separable from each other.

We trained the 4-3-1 feedforward network depicted in Fig. 5.1, using the backpropagation algorithm; the values $v = -1$, 0, and 1 were used for the classes Versicolor, Virginica, and Setosa, respectively. The output f was used to classify the input: if $f < -0.5$ then Versicolor; if $f \in [-0.5, 0.5]$ then Virginica; and if $f > 0.5$ then Setosa.

---

[1] The data is available online at the UCI Machine Learning Repository [6].

E. Kolman, M. Margaliot: Knowledge-Based Neurocomputing, STUDFUZZ 234, pp. 41–57.
springerlink.com © Springer-Verlag Berlin Heidelberg 2009

Knowledge Extraction Using the FARB

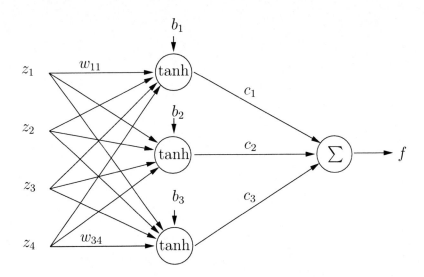

Fig. 5.1. ANN for the Iris classification problem

The trained network, with parameters[2]

$$W := \{w_{ij}\} = \begin{bmatrix} -0.4 & -5 & -0.3 & 0.7 \\ 150 & 150 & -67 & -44 \\ -5 & 9 & -7 & 2 \end{bmatrix}, \quad \mathbf{b} = \begin{bmatrix} -7 \\ -520 \\ -11 \end{bmatrix}, \quad \text{and } \mathbf{c} = \begin{bmatrix} -0.5 \\ 0.5 \\ -1 \end{bmatrix},$$

classifies the data set with 99% accuracy (149 correct classifications out of the 150 samples).

As in the previous examples, the network architecture and list of parameter values do not provide any explanation of its functioning. It is possible, of course, to express the network output as a mathematical formula

$$f = \sum_{j=1}^{3} c_j \tanh(\sum_{i=1}^{3} w_{ji} z_i + b_j),$$

but this does not seem to provide much intuition either. Hence, a more comprehensible representation is needed.

Let y_i, $i \in [1:3]$, denote the hidden neuron inputs, that is, $y_i := (\mathbf{w}^i)^T \mathbf{z}$, where $\mathbf{z} := (z_1, z_2, z_3, z_4)^T$, and $(\mathbf{w}^i)^T$ is the ith row of the matrix W. Corollary 3.6 implies that the feedforward ANN is equivalent to an eight-rule FARB with: $a_0 = 0$, $a_i = c_i$, $v_i = -b_i$ and $x_i = y_i$, $i = [1:3]$, that is:

R_1: If x_1 is st 7 and x_2 is st 520 and x_3 is st 11, Then $f = 1$,
R_2: If x_1 is st 7 and x_2 is st 520 and x_3 is lt 11, Then $f = -1$,

---

[2] The numerical values were rounded to one decimal digit, without affecting the accuracy of the classification.

R_3: If x_1 is *st* 7 and x_2 is *lt* 520 and x_3 is *st* 11, Then $f = 2$,
R_4: If x_1 is *st* 7 and x_2 is *lt* 520 and x_3 is *lt* 11, Then $f = 0$,
R_5: If x_1 is *lt* 7 and x_2 is *st* 520 and x_3 is *st* 11, Then $f = 0$,
R_6: If x_1 is *lt* 7 and x_2 is *st* 520 and x_3 is *lt* 11, Then $f = -2$,
R_7: If x_1 is *lt* 7 and x_2 is *lt* 520 and x_3 is *st* 11, Then $f = 1$,
R_8: If x_1 is *lt* 7 and x_2 is *lt* 520 and x_3 is *lt* 11, Then $f = -1$,

where $\{lt, st\}$ stands for $\{larger\ than,\ smaller\ than\}$, respectively, defined as in (2.7) with $\alpha = 2$.

These rules provide a complete *symbolic* description of the network operation. In order to increase the FARB comprehensibility, we apply the step-by-step simplification procedure described in Section 4.2.

STEP 1. Here, $a_1 = -0.5$, $a_2 = 0.5$, and $a_3 = -1$, so all the a_is are of the same magnitude, and none of them may be deleted.

STEP 2. A calculation yields $\{m_k l_k\}_{k=1}^8 = \{3, 3, 7e-6, 2, 1e-4, 2e-8, 6e-18, 3e-12\}$. Hence, we delete rules 3,5,6,7, and 8. We are left with three rules. Renumbering these rules as R_1 to R_3 yields:

R_1: If x_1 is *st* 7 and x_2 is *st* 520 and x_3 is *st* 11, Then $f = 1$,
R_2: If x_1 is *st* 7 and x_2 is *st* 520 and x_3 is *lt* 11, Then $f = -1$,
R_3: If x_1 is *st* 7 and x_2 is *lt* 520 and x_3 is *lt* 11, Then $f = 0$.

STEP 3. A calculation yields $e/r = 0.12$ (in fact, $e/r \leq 0.01$ for 97% of the examples in the training set), so we replace the COG defuzzifier with the MOM defuzzifier.

STEP 4. All three rules have the same first atom and, therefore, this atom is deleted.

STEP 5. The atom 'x_3 is *st* 11' appears only in the first rule, and the opposite atom 'x_3 is *lt* 11' appears in the other rules. Hence, the simplification yields

If x_3 is *st* 11, Then Class is Setosa;
Else, If x_2 is *st* 520, Then Class is Versicolor;
If x_2 is *lt* 520, Then Class is Virginica

(where we also replaced the numeric values in the Then-part with the names of the corresponding classes).

STEP 6. Defining the class Virginica as the default class yields

If x_3 is *st* 11, Then Class is Setosa;
Else, If x_2 is *st* 520, Then Class is Versicolor;
Else, Class is Virginica.

Note that the rule-base is now hierarchical: every rule is either valid (and then the corresponding class is immediately determined) or invalid (and then the next rule is examined). Hence, we can consider the truth values as belonging to $\{0, 1\}$ rather than $[0, 1]$. To do that, we set $t_j(\mathbf{x})$ to one (zero) if $t_j(\mathbf{x}) \geq 0.9$ (otherwise). This yields

44   Knowledge Extraction Using the FARB

If ($x_3 < 9$), Then Class is Setosa;
Else, If ($x_2 < 518$), Then Class is Versicolor;
        Else, Class is Virginica.

The classification accuracy of this simple rule-base is still identical to that of the trained ANN, that is, 99%.

STEP 7. The attribute values satisfy $z_1 \in [0.1, 2.5]$, $z_2 \in [1.0, 6.9]$, $z_3 \in [2.0, 4.4]$, and $z_4 \in [4.3, 7.9]$. Denoting the lower (upper) value of z_i by z_i^{min} (z_i^{max}), and defining normalized variables $\tilde{z}_i := (z_i - z_i^{min})/(z_i^{max} - z_i^{min})$, yields

$$x_2 = 360\tilde{z}_1 + 885\tilde{z}_2 - 160.8\tilde{z}_3 - 158.4\tilde{z}_4 - 158.2,$$

and

$$x_3 = -12\tilde{z}_1 + 53.1\tilde{z}_2 - 16.8\tilde{z}_3 + 7.2\tilde{z}_4 + 3.1.$$

Calculating over the training set, we find that

$$\frac{1}{150}\sum_{\mathbf{z} \in D}(-160.8\tilde{z}_3 - 158.4\tilde{z}_4 - 158.2) = -296.7,$$

and

$$\frac{1}{150}\sum_{\mathbf{z} \in D}(-12\tilde{z}_1 - 16.8\tilde{z}_3 + 7.2\tilde{z}_4 + 3.1) = -6.7.$$

Hence, we replace x_3 in the first rule by $53.1\tilde{z}_2 - 6.7$ and x_2 in the second rule by $360\tilde{z}_1 + 885\tilde{z}_2 - 296.7$. Restating our rule-base, using the original attribute names, yields the final rule-base:

If (Petal length < 2.75cm), Then Class is Setosa;
Else, If (Petal length + Petal width < 6.53cm),
        Then Class is Versicolor;
        Else, Class is Virginica.

This set of rules has a classification accuracy of 97% (146 out of 150 samples). Thus, STEP 7 considerably increased the interpretability of the rule-base at the cost of a 2% accuracy loss.[3]   □

Summarizing, transforming the ANN into an equivalent FARB, and simplifying the set of rules, led to a highly comprehensible description of the knowledge embedded in the trained network.

The Iris classification problem can be solved by a relatively small ANN. In the following section, we demonstrate the efficiency of our approach using a larger-scale ANN.

5.2   The LED Display Recognition Problem

Consider the problem of learning to identify the digit produced by a seven-segment *light emitting diodes* (LED) display [20]. This is a well-known problem

---

[3] For other classification rules obtained for this problem, see, e.g., [12; 24; 36; 37; 176; 191].

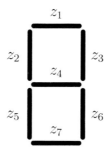

Fig. 5.2. LED display

in machine learning [101], and several pattern recognition algorithms have been applied in order to solve it, including classification trees [20], instance-based learning algorithms [1], ANNs [18], evolutionary strategies [13; 100], and support vector machines [13]. This problem was also used to demonstrate the performance of algorithms for noise rejection [101], dimensionality reduction [65], and knowledge acquisition [56].

The input to the learning algorithm is a set of supervised examples in the form $(z_1, z_2, \ldots, z_{24}, v)$. Here, $z_1, \ldots, z_7$ are the states of the seven diodes (1 = ON, 0 = OFF) of the LED display (see Fig. 5.2). For example, the vector $(1, 1, 0, 1, 1, 1)$ represents the digit 6, and $(1, 1, 1, 1, 0, 1, 1)$ represents the digit 9. The value $v \in [0:9]$ is the displayed digit. The inputs $z_8, \ldots, z_{24}$ are independent random variables with $\text{Prob}(z_i = 0) = \text{Prob}(z_i = 1) = 1/2$. These noise inputs make the problem more challenging, as the classification algorithm must also learn to discriminate between the meaningful and the useless inputs.

A trial and error approach suggested that the minimal number of hidden neurons needed for solving the LED recognition problem is six, so we used a 24-6-10 ANN.[4] The hidden neurons employ the hyperbolic tangent activation function, and the ten output neurons are linear. Preprocessing included converting the inputs from $\{0, 1\}$ to $\{-1, 1\}$.

The network was trained using a set of 2050 supervised examples.[5] Each of the ten outputs $f_0, \ldots, f_9$ corresponds to a different digit and the classification is based on the winner-takes-all approach [59]. After training, the ANN correctly classified all the examples. However, its architecture and parameter values do not provide any insight into the ANN functioning.

The ANN contains 220 parameters (204 weights and 16 biases), and extracting the embedded knowledge is a challenging task. Note that other fuzzy rule extraction methods were applied to smaller ANNs. In [97], fuzzy rules were extracted from a fuzzy neural network with up to 10 neurons and 60 parameters; in [129], interpretable fuzzy models were extracted from two neuro-fuzzy networks, with

---

[4] A similar network was used in [18]. Recently, a smaller ANN was successfully used in [157].
[5] For more details on the training process and the ANN parameters, see Appendix B.

46   Knowledge Extraction Using the FARB

28 and 92 parameters; and in [7; 66], fuzzy rules were extracted from a neural network with 36 parameters.

5.2.1 Knowledge Extraction Using the FARB

Let w_{ji} denote the weight connecting the input z_i, $i \in [1:24]$, to the jth hidden neuron, $j \in [1:6]$. The net input of the jth hidden neuron is then $y_j := \sum_{i=1}^{24} w_{ji} z_i$. Applying Corollary 3.6 transforms the ANN into an equivalent FARB with inputs $x_j = y_j$, $j \in [1:6]$, $2^6 = 64$ fuzzy rules, and output $\mathbf{f} = (f_0, \ldots, f_9)^T$. For example, Rule 53 in this FARB is:[6]

R_{53}: If $x_1 \approx 1$ & $x_2 \approx 1$ & $x_3 \approx -1$ & $x_4 \approx 1$ & $x_5 \approx -1$ & $x_6 \approx 1$,
      Then $\mathbf{f}^{53} = (-1.4, -2.5, -0.6, -0.7, -0.2, -0.5, -11, -1.4, 0, 0.4)^T$,

where & denotes "and", $\mathbf{f}^i \in \mathbb{R}^{10}$ is the value in the Then-part of rule i, and $x_i \approx k$ is shorthand for 'x_i equals k'. The terms {equals 1, equals -1} are modeled as in (2.1) with $\sigma^2 = 1$. The inferencing amounts to computing a weighted sum, $\mathbf{f} = (f_0, \ldots, f_9)^T$, of the 64 vectors in the Then-part of the rules, and the final classification is $i := \arg \max_{l \in [0:9]} \{f_l\}$.

This rule-set provides a *complete* symbolic representation of the ANN functioning, so we immediately obtain a fuzzy classifier that solves the LED recognition problem. However, its comprehensibility is hindered by the large number and the complexity of the rules. To gain more insight, we simplify this FARB.

5.2.2 FARB Simplification

We apply the step-by-step simplification procedure described in Section 4.2.

STEP 1. None of the a_is may be deleted.

STEP 2. We iteratively apply a slightly modified version of this step, which is more suitable for classification problems. Consider a FARB with input $\mathbf{x} \in \mathbb{R}^m$, $q := 2^m$ rules, and output $\mathbf{f} \in \mathbb{R}^l$. Let $t^i(\mathbf{x})$ denote the DOF of rule i. Inferring yields $\mathbf{f}(\mathbf{x}) = \mathbf{u}(\mathbf{x}) / \sum_{i=1}^{q} t^i(\mathbf{x})$, where $\mathbf{u}(\mathbf{x}) := \sum_{i=1}^{q} t^i(\mathbf{x}) \mathbf{f}^i$.

Modifying $t^k(\mathbf{x})$ to $\hat{t}^k(\mathbf{x})$ yields the modified output:

$$\hat{\mathbf{f}}(\mathbf{x}) = \frac{\hat{t}^k(\mathbf{x}) \mathbf{f}^k + \sum_{\substack{i \in [1:q] \\ i \neq k}} t^i(\mathbf{x}) \mathbf{f}^i}{\hat{t}^k(\mathbf{x}) + \sum_{\substack{i \in [1:q] \\ i \neq k}} t^i(\mathbf{x})}.$$

The classification decision will not change as long as $\arg \max_{i \in [1:l]} \{\hat{\mathbf{f}}_i(\mathbf{x})\} = \arg \max_{i \in [1:l]} \{\mathbf{f}_i(\mathbf{x})\}$, that is, if $\arg \max_{i \in [1:l]} \{\mathbf{u}_i(\mathbf{x}) - (t^k(\mathbf{x}) - \hat{t}^k(\mathbf{x})) \mathbf{f}_i^k\} = \arg \max_{i \in [1:l]} \{\mathbf{u}_i(\mathbf{x})\}$.[7]

---

[6] The numerical values were rounded to one decimal digit, without affecting the accuracy of the classification.

[7] Recall that deleting rule k from the FARB altogether amounts to setting $\hat{t}^k(\mathbf{x}) \equiv 0$.

$$
\begin{array}{l}
\text{Set } R(l,j) = 1, \quad \text{for all } l, j \\
\text{While (there is no index } l \text{ such that } R(l,j) = 0, \text{ for all } j) \\
\quad \text{For } j = 0 \text{ to } 9 \\
\qquad Q \leftarrow \{k | R(k,j) = 1\} \text{ /* rules in } Q \text{ are significant for digit } j \text{ */} \\
\qquad q \leftarrow \arg \min_{k \in Q} p_j^k \\
\qquad R(q,j) \leftarrow 0 \text{ /* mark rule } q \text{ as insignificant for digit } j \text{ */} \\
\quad \text{EndFor} \\
\text{EndWhile} \\
\text{Output}(l) \text{ /* rule } l \text{ is insignificant for all the digits */}
\end{array}
$$

Fig. 5.3. Simplification procedure

Let D denote the training set, and let $(\mathbf{f}^k)_j$ be the jth element of the output vector in rule k. It follows from the analysis above that if $p_j^k := \max_{\mathbf{x} \in D} t^k(\mathbf{x})$ $(\mathbf{f}^k)_j$ is small, then the kth rule has a small positive effect on the classification of digit j. This suggests the procedure depicted in Fig. 5.3. This procedure calculates a matrix $R(\cdot, \cdot)$ such that $R(l,j) = 1$ ($R(l,j) = 0$) indicates that rule l is "significant" ("insignificant") for the correct classification of digit j.

The procedure output is an index l indicating that rule l has a small effect on classifying *all* the ten digits. Then, rule l is deleted if removing it from the rule-base does not change the classification for all the training examples.

Applying this procedure repeatedly to our rule-base leads to the deletion of 54 out of the 64 rules. The result is a simplified FRB with only ten rules that correctly classifies the training set.

STEPS 3-6. The conditions for applying these steps are not satisfied.

STEP 7. Examining the w_{ji}s yields

$$
\frac{\min_{i \leq 7} |w_{ji}|}{\max_{i \geq 8} |w_{ji}|} > 240, \qquad \text{for all } j \in [1:6].
$$

Thus, we replace the term $w_{ji}z_i$ with the term $w_{ji}\bar{z}_i = w_{ji}/2$, for all $i \in [8:24]$, and all $j \in [1:6]$.

At this point, each x_j in the ten-rule FRB is a linear combination of only seven inputs: $z_1, \ldots, z_7$. The noise inputs $z_8, \ldots, z_{24}$ were correctly identified as meaningless. Of course, this step is identical to setting weak connections in the ANN to zero. However, note that the symbolic structure of the FARB also allowed us to perform simplification steps that cannot be carried out in terms of the weights of the ANN.

Evidently, the FRB obtained at this point is much simpler than the original one, as the number of rules is reduced from 64 to 10, and the number of antecedents in each rule is reduced from 24 to 7 (see Table 5.1). Furthermore, this FRB is simple enough to allow us to interpret its functioning.

48 Knowledge Extraction Using the FARB

Table 5.1. The ten fuzzy rules. R denotes the rule number in the final FRB, $x_i = 1$ ($x_i = -1$) implies that the rule includes the term 'x_i equals 1' ('x_i equals -1') in its If-part. $\mathbf{f}$ is the rule output vector. Bold font denotes the largest element in each vector.

R	x_1	x_2	x_3	x_4	x_5	x_6	$\mathbf{f}$
0	-1	-1	-1	-1	1	1	$(\mathbf{1.3},0.1,-1.6,-2,-1.4,-0.1,-0.8,-1.2,-1.2,-0.9)$
1	-1	-1	-1	-1	1	-1	$(-0.1,\mathbf{0.9},-1.6,-1.1,-1.5,-0.9,0.2,0,-1.5,-2.3)$
2	1	-1	1	-1	-1	1	$(-1.1,0.2,\mathbf{0.6},-1.5,-0.1,-1.3,-1.7,-0.5,-1.5,-1.1)$
3	1	1	1	-1	1	-1	$(-1.4,-1.1,-0.5,\mathbf{1.4},-1.4,-1.6,-0.8,-1.4,-0.2,-1.1)$
4	1	1	1	1	-1	1	$(-2.3,-2,-0.1,-0.4,\mathbf{0},-0.1,-0.7,-1.5,-0.8,-0.1)$
5	-1	1	1	1	1	1	$(-0.7,-1.7,-1.8,-0.7,-1,\mathbf{1.3},0.6,-2.2,-1.3,-0.5)$
6	-1	1	1	-1	-1	1	$(-0.6,-0.5,-0.8,-2.8,-0.1,-0.4,\mathbf{1},-1.5,-0.5,-1.7)$
7	1	-1	-1	1	-1	-1	$(-1.5,-0.8,-1.1,-0.8,-0.7,-1.5,-1.4,\mathbf{1.1},-0.6,-0.7)$
8	1	1	-1	-1	-1	-1	$(-1.2,-1.3,-0.6,-0.6,-0.7,-2.6,-0.7,-0.3,\mathbf{1},-1.2)$
9	1	-1	-1	1	1	1	$(-0.3,-1.4,-0.9,0.4,-1.3,0,-2.4,-1.2,-1.5,\mathbf{0.7})$

5.2.3 Analysis of the FRB

The symbolic structure of FRBs makes them much easier to understand than ANNs. In particular, we can analyze the operation of an FRB by examining the If-part and the Then-part of each rule.

The If-part

To understand the If-part of the rules, consider their DOF for each possible input, i.e., the 128 possible binary vectors $(z_1, \ldots, z_7)$. The ratio between the highest DOF and the second highest DOF for the ten rules is: 9.3, 12.7, 3.4, 1.5, 5.4, 5.4, 4.5, 2.3, 19.4, and 2.4, respectively. Thus, with the exception of R_3 (recall that the rules in our example are numbered R_0 to R_9), every rule is tuned to a specific input pattern and yields a much lower DOF for any other pattern.

Fig. 5.4 depicts the pattern yielding the highest DOF for each rule. It may be seen that rules R_1, R_5, R_6, and R_8 are tuned to recognize the digits 1, 5, 6 and 8, respectively. Rules R_0 and R_7 are tuned to patterns that are one Hamming distance away from the real digits 0 and 7.

Comparing the DOFs for only the ten patterns representing the digits $[0:9]$ shows that R_2 and R_3 have the highest DOF when the input is the digit one, and R_4 has the highest DOF when the input is the digit five. For all other rules, R_i obtains the highest DOF when the input is digit i.

The Then-part

It is easy to verify that $\arg\max_k (\mathbf{f}^i)_k = i$ for all i. In other words, if only rule i fired, the FRB classification decision would be digit i. In most rules, there is a

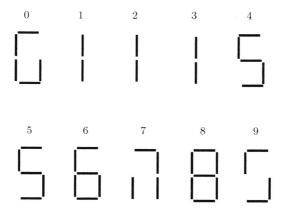

Fig. 5.4. Input pattern yielding maximal DOF

considerable difference between entry i and the second largest entry in $\mathbf{f}^i$. In five of the ten rules, the largest entry is positive while all other entries are negative. When such a rule fires it contributes not only to the classification as a specific digit, but also contributes negatively to all other possible classifications.

The analysis of the If- and Then-parts of the rules indicates that the FRB includes seven rules that are tuned to a specific digit: 0, 1, 5, 6, 7, 8, and 9. Each of these rules has a high DOF when the input is the appropriate digit. On the other hand, rules R_2, R_3, and R_4 are not tuned to the corresponding digit (for example, R_2 displays the highest DOF when the input is actually the digit one).[8] This behavior motivated us to try and understand the distinction between the two sets of digits:

$$S_1 := \{0, 1, 5, 6, 7, 8, 9\}, \quad \text{and} \quad S_2 := \{2, 3, 4\}. \tag{5.1}$$

Let $H(d_1, d_2)$ denote the Hamming distance between the LED representations of digits d_1 and d_2 (e.g., $H(1, 7) = 1$). Let M_i denote the set of digits with closest digit at distance i, i.e.,

$$M_i = \{d : \min_{\substack{j \in [0:9] \\ j \neq d}} H(d, j) = i\}.$$

Then,
$$M_1 = \{0, 1, 3, 5, 6, 7, 8, 9\}, \quad \text{and} \quad M_2 = \{2, 4\}. \tag{5.2}$$

Comparing (5.1) and (5.2) shows that there is a high correspondence between the sets M_i and S_i. Also, the definition of M_i suggests that digits in M_1 may be more difficult to recognize correctly than those in M_2. Thus, the FRB (or the original ANN) dedicates specially tuned rules for the more "tricky" digits.

---

[8] The rule-base correctly classifies all ten digits (including the digits 2, 3, and 4) because of the weighted combination of all the rules outputs.

50   Knowledge Extraction Using the FARB

The notion that digits that are more difficult to recognize deserve more attention is quite intuitive. However, understanding that the ANN implements this notion by observing its weights and biases is all but impossible. It is only through the knowledge extraction process that this notion emerges.

Summarizing, we demonstrated our knowledge extraction approach by applying it to an ANN trained to solve the LED recognition problem. The 24-6-10 network was transformed into a set of 64 fuzzy rules. Simplification of this ruleset led to a comprehensible representation of the ANN functioning. We showed, for example, that it is possible to conclude that the ANN learned to place an emphasis on the specific digits that are more difficult to recognize.

5.3   The L_4 Language Recognition Problem

We now describe an application of the ANN–FARB equivalence for knowledge extraction from an ANN trained to solve a language recognition problem. For the sake of completeness, we briefly review some ideas from the field of formal languages.

5.3.1   Formal Languages

Let Σ denote some set of symbols (e.g., $\Sigma = \{a, b, c, \ldots, z\}$ or $\Sigma = \{0, 1\}$). A *string* is a finite-length sequence of symbols from Σ. Let $*$ denote the Kleene closure operator [84], so Σ^* is the (infinite) set of all the strings constructed over Σ.

Definition 5.1. [62] *A formal language is a set of strings $L \subseteq \Sigma^*$.*

A formal language can be naturally associated with the grammar that generates the language.

Definition 5.2. *A formal grammar is a quadruple $G = \langle S, N, T, P \rangle$, where S is the start symbol, N and T are non-terminal and terminal symbols, respectively, and P is a set of production rules of the form $u \to v$, where $u, v \in (N \cup T)^*$, and u contains at least one non-terminal symbol.*

Note that repeatedly applying the production rules generates a specific set of strings, that is, a language. This language is denoted $L(G)$.

Chomsky and Schützenberger [25; 26] sorted formal languages into four types: *recursive, context-sensitive, context-free,* and *regular*. Each class is strictly contained in its predecessor (e.g., the set of regular languages is strictly contained in the set of context-free languages). The classes are defined by the type of production rules allowed in the grammar. Regular languages, generated by regular grammars, represent the smallest class in the hierarchy of formal languages.

Definition 5.3. [128] *A regular grammar G is a formal grammar with production rules of the form $A \to a$ or $A \to aB$, where $A, B \in N$ and $a \in T$.*

The L_4 Language Recognition Problem   51

Example 5.4. Tomita's 4th grammar [171] is the regular grammar defined by:

S is the start symbol,
$N = \{S, A, B\}$,
$T = \{0, 1\}$,
$P = \{S \rightarrow 1S, S \rightarrow 0A, A \rightarrow 1S, A \rightarrow 0B, B \rightarrow 1S, S \rightarrow \varepsilon,$
$\quad A \rightarrow \varepsilon, B \rightarrow \varepsilon \}$,

where ϵ denotes the empty string. This grammar generates a regular language, denoted L_4, which is the set of all binary strings that do *not* include '000' as a substring. □

It is natural to associate a regular language with a *deterministic finite-state automaton* (DFA) that can distinguish between strings that belong/do not belong to the language.

Definition 5.5. *A DFA is a 5-tuple $M = \langle \Sigma, Q, R, F, \delta \rangle$, where Σ is the alphabet, $Q = \{s_1, \ldots, s_m\}$ is a set of states, $R \in Q$ is the start state, $F \subseteq Q$ is a set of accepting states, and $\delta : Q \times \Sigma \rightarrow Q$ defines the state transitions.*

We say that a string x is *accepted* by a DFA M iff $s(x)$, the state that is reached after x has been read by M, is an accepting state [126]. For each regular language $L(G)$, there is an associated DFA M, such that a string x is accepted by M iff $x \in L(G)$.

5.3.2 Formal Languages and RNNs

RNNs are often used for recognizing formal languages. The training is performed using a set containing pairs of the form {string, label}, where label indicates whether the string belongs to the language or not. KE techniques are commonly demonstrated by applying them to RNNs that have learned to recognize formal languages [5; 50; 113].

Usually, the knowledge embedded in the RNN is extracted in the form of a DFA, that represents the underlying regular grammar. For example, Giles *et al.* [50; 124] trained RNNs with 3 to 5 hidden neurons to recognize L_4. Figure 5.5 depicts a four-state DFA extracted from one of those networks [124]. It is easy to see that this DFA indeed correctly recognizes L_4. Some disadvantages of this KE approach were described in Section 1.4.1 above.

To demonstrate the efficiency of KE based on the FARB–ANN equivalence, we trained an RNN to recognize L_4, and then extracted the knowledge embedded within the RNN as a FARB.

5.3.3 The Trained RNN

The architecture included three hidden neurons, one input neuron and one bias neuron. The network is thus described by: $s_0(t) \equiv 1$,

$$s_i(t) = \sigma \left(\sum_{j=0}^{4} w_{ij} s_j(t-1) \right), \qquad i = 1, 2, 3, \tag{5.3}$$

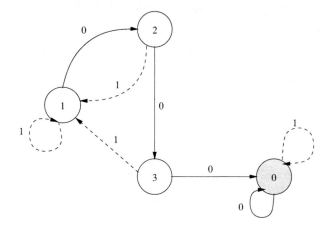

Fig. 5.5. A DFA that recognizes L_4. State 1 is the starting state, states 1 – 3 are accepting states, and state 0 is a rejecting state. Arrows indicate state transitions.

with the input neuron

$$s_4(t) = \begin{cases} 1, & \text{if } I(t) = 0, \\ -1, & \text{if } I(t) = 1, \end{cases} \tag{5.4}$$

where $I(t)$ is the value of the input at time t.

The RNN was trained using *real time recurrent learning* (RTRL) [183], with learning rate $\eta = 0.1$, momentum $\gamma = 0.05$, and an added regularization term $10^{-3} \sum_{i,j} w_{ij}^2$ [69].

We generated 2000 RNNs with initial weights and biases drawn from a uniform probability distribution over $[-2, 2]$ and $[-1, 1]$, respectively. Each RNN was trained for 700 epochs, where each epoch consisted of presenting the full training set to the RNN, and updating the weights after each presented string.

The data set contained all the 8184 binary strings with length $3 \leq l \leq 12$ (of which 54% do not belong to L_4, i.e., include '000' as a substring), and was split into training and test sets. The training set included all the 248 binary strings with length $3 \leq l \leq 7$ (of which 32% do not belong to L_4), and the test set included the 7936 binary strings with length $8 \leq l \leq 12$ (of which 55% do not belong to L_4).[9] Our main focus is on extracting the knowledge from the trained network, and not on finding the optimal RNN for the given task. Therefore, we used the simple holdout approach rather than more sophisticated and time consuming approaches, such as bootstrap or cross-validation [14; 85].

---

[9] Since the performance of the RTRL algorithm deteriorates for longer strings, it is common to use shorter strings for training and then test the generalization capability of the trained network on longer strings.

After training, the RNN with the best performance had the following parameters:[10]

$$W = \begin{bmatrix} -7.6 & 15.2 & 8.4 & 0.2 & 3 \\ -4.7 & 0.2 & -0.2 & 4.5 & 1.5 \\ -0.2 & 0 & 0 & 0 & 4 \end{bmatrix}. \tag{5.5}$$

This RNN correctly classifies all the strings in the data set. Furthermore, the following result establishes the correctness of the RNN.

Proposition 5.6. *The RNN defined by (5.3), (5.4), and (5.5) correctly classifies any given binary string according to L_4.*

PROOF. See Appendix A.

Although this RNN functions quite well, it is very difficult to understand exactly what it does. To gain more understanding, we represent the dynamic behavior of each neuron as an equivalent FARB, and then simplify the FARB.

5.3.4 Knowledge Extraction Using the FARB

Extracting rules for $s_3(t)$

The dynamic equation for $s_3(t)$ is

$$s_3(t) = \sigma \left(4s_4(t-1) - 0.2\right). \tag{5.6}$$

Corollary 3.3 implies that this is equivalent to the FARB:

R_1: If $2s_4(t-1)$ is *larger than* 0.1, Then $s_3(t) = 1$,
R_2: If $2s_4(t-1)$ is *smaller than* 0.1, Then $s_3(t) = 0$,

where the fuzzy terms {*larger than, smaller than*} are modeled (from hereon) using the Logistic functions (2.7) with $\alpha = 2$.

Rule Simplification

We now apply the simplification procedure described in Section 4.2 to the two-rule FARB above.

STEP 1. The FARB parameters are $a_0 = a_1 = 1/2$, so none of the a_is may be deleted.
STEP 2. $m_k l_k = 0.98$ for $k = 1, 2$, so none of the rules may be deleted.
STEP 3. $e/r = 0.0219$, so we replace the COG defuzzifier with the MOM defuzzifier. Since $s_4(t)$ can attain only two values, this yields the *crisp* rules:

R_1: If $s_4(t-1) = 1$, Then $s_3(t) = 1$,
R_2: If $s_4(t-1) = -1$, Then $s_3(t) = 0$.

---

[10] The numerical values were rounded to one decimal digit, without affecting the accuracy of the classification.

Restating this in terms of the input yields:

R_1: If $I(t-1) = 0$, Then $s_3(t) = 1$,
R_2: If $I(t-1) = 1$, Then $s_3(t) = 0$.

STEP 4. There is no atom that appears in both rules.
STEP 5. Applying this step does not affect the rules.
STEP 6. Setting the case $s_3(t) = 0$ as the default class yields

$$\text{If } I(t-1) = 0, \text{ Then } s_3(t) = 1; \text{ Else, } s_3(t) = 0.$$

This rule provides a clear interpretation of the functioning of this neuron, namely, $s_3(t)$ is ON (i.e., equals 1) iff the last input bit was '0'.

Extracting rules for $s_2(t)$

The dynamic equation for $s_2(t)$ is

$$s_2(t) = \sigma\big(0.2s_1(t-1) - 0.2s_2(t-1) + 4.5s_3(t-1) + 1.5s_4(t-1) - 4.7\big).$$

Replacing the relatively small variables with their expectation yields[11]

$$s_2(t) = \sigma\left(4.5s_3(t-1) + 1.5s_4(t-1) - 4.7\right) + e_1\left(s_1(t-1), s_2(t-1)\right), \quad (5.7)$$

where $|e_1\left(s_1(t-1), s_2(t-1)\right)| \leq 0.04$. Neglecting e_1 has no effect on the network classification results over the test set. That is, the simplified RNN correctly classifies all the examples in the test set. Substituting (5.6) in (5.7) and neglecting e_1 yields

$$\sigma^{-1}\left(s_2(t)\right) = 4.5\sigma\left(4s_4(t-2) - 0.2\right) + 1.5s_4(t-1) - 4.7. \quad (5.8)$$

The right hand side of this equation is not a sum of sigmoid functions, and thus cannot be interpreted using a suitable FARB. To overcome this, note that if $v \in \{-1, 1\}$ then $v = 2\sigma(5v) - 1 + e_2$, with $|e_2| \leq 0.014$. We can thus write:

$$s_4(t-1) = 2\sigma\left(5s_4(t-1)\right) - 1 + e_2. \quad (5.9)$$

Substituting this in (5.8) and neglecting e_2 yields

$$\sigma^{-1}\left(s_2(t)\right) = 4.5\sigma\left(4s_4(t-2) - 0.2\right) + 1.5\left(2\sigma\left(5s_4(t-1)\right) - 1\right) - 4.7$$
$$= 4.5\sigma\left(4s_4(t-2) - 0.2\right) + 3\sigma\left(5s_4(t-1)\right) - 6.2.$$

---

[11] This is equivalent to applying Step 7 of the simplification procedure to the corresponding FARB.

By Corollary 3.5, this is equivalent to the four-rule FARB:

R_1: If $2s_4(t-2)$ is *larger than* 0.1 and $2.5s_4(t-1)$ is *larger than* 0,
  Then $\sigma^{-1}(s_2(t)) = 1.3$,
R_2: If $2s_4(t-2)$ is *larger than* 0.1 and $2.5s_4(t-1)$ is *smaller than* 0,
  Then $\sigma^{-1}(s_2(t)) = -1.7$,
R_3: If $2s_4(t-2)$ is *smaller than* 0.1 and $2.5s_4(t-1)$ is *larger than* 0,
  Then $\sigma^{-1}(s_2(t)) = -3.2$,
R_4: If $2s_4(t-2)$ is *smaller than* 0.1 and $2.5s_4(t-1)$ is *smaller than* 0,
  Then $\sigma^{-1}(s_2(t)) = -6.2$.

The next step is to simplify this FARB.

Rule Simplification

STEP 1. The FARB parameters are $a_0 = -2.45$, $a_1 = 2.25$, and $a_2 = 1.5$, so none of the a_is may be deleted.
STEP 2. $\{m_k l_k\}_{k=1}^4 = \{7.29, 4.37, 4.4, 7.34\}$, so no rule may be deleted.
STEP 3. $e/r = 0.16$, so we replace the COG defuzzifier with the MOM defuzzifier. Using (5.4) yields the crisp rules:

R_1: If $I(t-2) = 0$ and $I(t-1) = 0$, Then $s_2(t) = \sigma(1.3) \approx 0.8$,
R_2: If $I(t-2) = 0$ and $I(t-1) = 1$, Then $s_2(t) = \sigma(-1.7) \approx 0.15$,
R_3: If $I(t-2) = 1$ and $I(t-1) = 0$, Then $s_2(t) = \sigma(-3.2) \approx 0$,
R_4: If $I(t-2) = 1$ and $I(t-1) = 1$, Then $s_2(t) = \sigma(-6.2) \approx 0$.

STEPS 4,5. The conditions for applying these steps are not satisfied here.
STEP 6. Defining the output $s_2(t) = 0$ as the default value, and approximating the output of the second rule by $s_2(t) = 0$, yields

$$\text{If } I(t-2) = 0 \text{ and } I(t-1) = 0, \text{ Then } s_2(t) = 0.8; \text{ Else, } s_2(t) = 0.$$

It is now easy to understand the functioning of this neuron, namely, $s_2(t)$ is ON iff the last two input bits were '00'.

Extracting rules for $s_1(t)$

The dynamic equation for s_1 is

$$s_1(t) = \sigma\left(15.2s_1(t-1) + 8.4s_2(t-1) + 0.2s_3(t-1) + 3s_4(t-1) - 7.6\right).$$
$$(5.10)$$

There is one term with a relatively small weight: $0.2s_3(t-1)$. Using (5.6) yields

$$s_3(t-1) = \sigma\left(4s_4(t-2) - 0.2\right)$$
$$= \begin{cases} 0.98, & \text{if } s_4(t-2) = 1, \\ 0.01, & \text{if } s_4(t-2) = -1. \end{cases}$$

Assuming a symmetric distribution

$$Prob\left(s_4(t-2) = 1\right) = Prob\left(s_4(t-2) = -1\right) = 1/2$$

56   Knowledge Extraction Using the FARB

yields $E\{0.2s_3(t-1)\} = 0.1$, so we simplify (5.10) to

$$s_1(t) = \sigma\left(15.2s_1(t-1) + 8.4s_2(t-1) + 3s_4(t-1) - 7.5\right). \tag{5.11}$$

This approximation does not affect the network classification results on the test set. Combining (5.11) with (5.6),(5.7), and (5.9) yields

$$\begin{aligned}
\sigma^{-1}\left(s_1(t)\right) =& 15.2s_1(t-1) + 8.4s_2(t-1) + 3s_4(t-1) - 7.5 \\
=& 15.2s_1(t-1) \\
& + 8.4\sigma\left(4.5\sigma\left(4s_4(t-3) - 0.2\right) + 1.5s_4(t-2) - 4.7\right) \\
& + 6\sigma\left(5s_4(t-1)\right) - 10.5. \tag{5.12}
\end{aligned}$$

Since

$$\sigma\left(4s_4(t-3) - 0.2\right) = 0.48s_4(t-3) + 0.5 + e_3\left(s_4(t-3)\right),$$

with $|e_3\left(s_4(t-3)\right)| < 0.005$, Eq. (5.12) can be simplified to

$$\begin{aligned}
\sigma^{-1}\left(s_1(t)\right) =& 8.4\sigma\left(2.16s_4(t-3) + 1.5s_4(t-2) - 2.45\right) \\
& + 6\sigma\left(5s_4(t-1)\right) + 15.2s_1(t-1) - 10.5. \tag{5.13}
\end{aligned}$$

Applying Corollary 3.5 to (5.13) yields the FARB:

R_1: If $1.08s_4(t-3) + 0.75s_4(t-2)$ is lt 1.23 and $2.5s_4(t-1)$ is lt 0,
    Then $\sigma^{-1}(s_1(t)) = 3.9 + 15.2s_1(t-1)$,
R_2: If $1.08s_4(t-3) + 0.75s_4(t-2)$ is lt 1.23 and $2.5s_4(t-1)$ is st 0,
    Then $\sigma^{-1}(s_1(t)) = -2.1 + 15.2s_1(t-1)$,
R_3: If $1.08s_4(t-3) + 0.75s_4(t-2)$ is st 1.23 and $2.5s_4(t-1)$ is lt 0,
    Then $\sigma^{-1}(s_1(t)) = -4.5 + 15.2s_1(t-1)$,
R_4: If $1.08s_4(t-3) + 0.75s_4(t-2)$ is st 1.23 and $2.5s_4(t-1)$ is st 0,
    Then $\sigma^{-1}(s_1(t)) = -10.5 + 15.2s_1(t-1)$,

where lt (st) stands for *larger than* (*smaller than*).

Rule Simplification

STEP 1. The FARB parameters are $a_0 = -3.3 + 15.2s_1(t-1)$, $a_1 = 4.2$, $a_2 = 3$, so none of the a_is may be deleted.
STEP 2. $\{m_k l_k\}_{k=1}^4 = \{10.99, 6.41, 8.33, 14.27\}$, so no rule may be deleted.
STEPS 3-6. The conditions for applying these steps are not satisfied. Specifically, $e/r = 3.41$, so the MOM inferencing approximation cannot be used.
    Restating the fuzzy rules in terms of the input yields:

R_1: If $2.16I(t-3) + 1.5I(t-2)$ is st 0.6 and $5I(t-1)$ is st 2.5,
    Then $\sigma^{-1}(s_1(t)) = f_1$,
R_2: If $2.16I(t-3) + 1.5I(t-2)$ is st 0.6 and $5I(t-1)$ is lt 2.5,
    Then $\sigma^{-1}(s_1(t)) = f_2$,
R_3: If $2.16I(t-3) + 1.5I(t-2)$ is lt 0.6 and $5I(t-1)$ is st 2.5,
    Then $\sigma^{-1}(s_1(t)) = f_3$,
R_4: If $2.16I(t-3) + 1.5I(t-2)$ is lt 0.6 and $5I(t-1)$ is lt 2.5,
    Then $\sigma^{-1}(s_1(t)) = f_4$,

where $\mathbf{f} = (f_1, f_2, f_3, f_4)^T = (3.9 + 15.2s_1(t-1), -2.1 + 15.2s_1(t-1), -4.5 + 15.2s_1(t-1), -10.5 + 15.2s_1(t-1))^T$. To better understand this FARB, we consider two extreme cases: $s_1(t-1) = 0.9$ and $s_1(t-1) = 0.1$.

- For $s_1(t-1) = 0.9$, the output is $\mathbf{f} = (17.6, 11.6, 9.2, 3.2)^T$. Since the inferencing yields a weighted sum of the f_is, with non-negative weights, $\sigma^{-1}(s_1(t)) \geq 3.2$, so $s_1(t) \geq 0.96$. Roughly speaking, this implies that if $s_1(t-1)$ is ON, then $s_1(t)$ is also ON, regardless of the input.
- For $s_1(t-1) = 0.1$, the output is $\mathbf{f} = (5.4, -0.6, -3, -9)^T$, and $\sigma(\mathbf{f}) = (1, 0.35, 0.05, 0)^T$. This implies that $\sigma(\mathbf{f}) > 0.5$ iff the first rule has a high DOF. Examining the If-part of the first rule, we see that this happens only if $I(t-1) = I(t-2) = I(t-3) = 0$. In other words, $s_1(t)$ will be turned ON only if the last three consecutive inputs were zero.

Recalling that the network rejects a string iff f_{out}–the value of s_1 after the entire string was fed into the network–is larger than 0.5, we can now easily explain the entire RNN functioning as follows. The value of neuron s_1 is initialized to OFF. It switches to ON whenever three consecutive zeros are encountered. Once it is ON, it remains ON, regardless of the input. Thus, the RNN recognizes strings containing a '000' substring.

Summarizing, in these three examples–the Iris classification problem, the LED recognition problem, and the L_4 recognition problem–the FARB–ANN equivalence was applied to extract fuzzy rules that describe the ANN behavior. Simplifying these symbolic rules provided a comprehensible explanation of the ANNs internal features and performance.

6 Knowledge-Based Design of ANNs

Corollaries 3.1-3.10 in Chapter 3 provide a transformation between a FARB and an ANN. The ANN type (i.e., feedforward, first-order RNN, or second-order RNN), structure, and parameter values, are determined directly from the FARB structure and parameters.

This suggests a novel scheme for knowledge-based design (KBD) of ANNs. Given the initial knowledge, determine the relevant inputs, denoted $x_1, \ldots, x_m$, and the number of outputs. For each output, restate the initial knowledge in the form of an FRB relating some subset of the inputs $\{y_1, \ldots, y_k\} \subseteq \{x_1, \ldots, x_m\}$ to this output. In this FRB, each y_i must be characterized using two fuzzy terms. The Then-part of each rule must be decomposed as a sum $a_0 \pm a_1 \pm \cdots \pm a_k$, where the signs are determined according to the If-part of the rule. More rules are added to the FRB, if necessary, until it contains 2^k fuzzy rules, expanding all the possible permutations of the input variables. The output of each added rule is again a linear sum of the a_is with appropriate signs. MFs for each input variable are chosen such that (2.3) holds.

At this point, the FRB becomes a FARB, and inferring yields an IO mapping that can be realized by an ANN. Thus, the initial knowledge on the problem domain was converted into an ANN.

The most delicate stage is finding suitable values for the parameters a_i, $i \in [0:k]$. If the number of fuzzy rules obtained from the initial knowledge is denoted by p, determining the a_is amounts to solving p equations in $k+1$ unknowns. If this set of equations does not have a solution, then the Then-part of the rules must be modified, without significantly altering the knowledge embedded in them. This can be done, for example, by adding small correction terms, that is changing the output from f to $f + \epsilon$, or by setting the rules outputs to the value $r(f)$, for some suitable function r.

In this chapter, we show how this general scheme can be used in practice. We present two novel KBD approaches. The *direct approach* follows the above scheme quite closely. This is somewhat ad-hoc and each design problem must be treated from scratch. In order to overcome these difficulties, we also introduce a *modular approach*. This is based on designing basic functional modules and

---

E. Kolman, M. Margaliot: Knowledge-Based Neurocomputing, STUDFUZZ 234, pp. 59–76.
springerlink.com © Springer-Verlag Berlin Heidelberg 2009

60   Knowledge-Based Design of ANNs

realizing them using a suitable ANN once and for all. Then, the KBD problem is addressed by suitably combining the basic modules.

The efficiency of these approaches is demonstrated by using them to design RNNs that solve formal recognition problems. These include the recognition of *non-regular* languages, so common KBD techniques, that are based on the DFA-to-RNN conversion, cannot be applied at all.

6.1   The Direct Approach

The direct approach is based on expressing the solution of the given problem using a suitable FRB, and then transforming it into a FARB. It is useful to decompose the original design problem into a subset of smaller and simpler sub-problems, and then design a FARB that solves each sub-problem. We now demonstrate how this works by designing an RNN that recognizes the L_4 language (see Section 5.3.1).

6.1.1   KBD of an ANN Recognizing L_4

The design is based on using four state variables: $s_1(t), \ldots, s_4(t)$. The basic idea is simple: $s_4(t)$, $s_3(t)$, and $s_2(t)$ will serve as memory cells that record the values of the input bits $I(t), I(t-1)$, and $I(t-2)$, respectively. The variable $s_1(t)$ will combine this information to indicate whether the substring '000' was detected or not. Thus, $s_1(t) = 1$ if either $s_1(t-1) = 1$ (i.e., a '000' substring was already detected previously) or if $s_4(t), s_3(t), s_2(t)$ indicate that the last three consecutive input digits were '000'. After feeding an input string x, with length l, $s_1(l+1)$ will be ON iff x contains the substring '000'.

We define $s_4(t)$ as in (5.4), i.e., $s_4(t)$ is ON iff the current input bit is zero. We state the required functioning of $s_i(t), i = 1, 2, 3$, in terms of a suitable FRB.

Designing a FARB for $s_3(t)$

The variable $s_3(t)$ should be ON (OFF) if $s_4(t-1) = 1$ ($s_4(t-1) = -1$). This can be stated as:

R_1: If $s_4(t-1)$ equals 1, Then $s_3(t) = 1$,
R_2: If $s_4(t-1)$ equals -1, Then $s_3(t) = 0$.

This is a two-rule FARB with $a_0 = a_1 = 1/2$. The terms {*equals* 1, *equals* -1} are modeled (from hereon) as in (2.1) with $\sigma^2 = 1$, i. e.

$$\mu_{=1}(y) := \exp\left(-\frac{(y-1)^2}{2}\right), \quad \text{and} \quad \mu_{=-1}(y) := \exp\left(-\frac{(y+1)^2}{2}\right).$$

Inferring yields

$$s_3(t) = \sigma\left(2s_4(t-1)\right). \tag{6.1}$$

Note that since $s_4(t) \in \{-1, 1\}$ for all t (see (5.4)),

$$s_3(t) \in \{\sigma(-2), \sigma(2)\} = \{0.12, 0.88\}, \quad \text{for all } t > 1.$$

The Direct Approach   61

Designing a FARB for $s_2(t)$

The variable $s_2(t)$ should be ON (OFF) if $s_4(t-2) = 1$ ($s_4(t-2) = -1$). This can be stated as:

R_1: If $s_4(t-2)$ equals 1, Then $s_2(t) = 1$,
R_2: If $s_4(t-2)$ equals -1, Then $s_2(t) = 0$.

This is a FARB, yet it is not suitable for our purposes, since the If-part should only include values at time $t-1$. To overcome this, we note that (6.1) yields

$$s_4(t-2) = \begin{cases} 1, & \text{if } s_3(t-1) = 0.88, \\ -1, & \text{if } s_3(t-1) = 0.12. \end{cases}$$

We therefore use the linear approximation $s_4(t-2) = 2.63s_3(t-1) - 1.31 + e$. Using this, and neglecting the error term e, yields the rules:

R_1: If $2.63s_3(t-1)$ equals 2.31, Then $s_2(t) = 1$,
R_2: If $2.63s_3(t-1)$ equals 0.31, Then $s_2(t) = 0$.

This is a FARB and inferring, with the terms {*equals 2.31, equals 0.31*} modeled using (2.1) with $\sigma^2 = 1$, yields:

$$s_2(t) = \sigma \left(5.26s_3(t-1) - 2.62\right). \tag{6.2}$$

Designing a FARB for $s_1(t)$

The variable $s_1(t)$ should be ON if either $s_4(t-3) = s_4(t-2) = s_4(t-1) = 1$, or if $s_1(t-1)$ is ON. This can be stated in a symbolic form as:

R_1: If $s_1(t-1)$ is *larger than* $1/2$, Then $s_1(t) = 1$,
R_2: If $s_4(t-3)$ equals 1 and $s_4(t-2)$ equals 1 and $s_4(t-1)$ equals 1,
    Then $s_1(t) = 1$,
R_3: Else, $s_1(t) = 0$.

We now modify this FRB in order to make it a FARB. In view of Corollary 3.6, we replace the output term $s_1(t) = 1$ [$s_1(t) = 0$] by $\sigma^{-1}(s_1(t)) \geq \alpha$ [$\sigma^{-1}(s_1(t)) \leq -\alpha$], where $\alpha \gg 0$.[1] Note that $s_1(t-1) \in [0,1]$, whereas $s_4(t) \in \{-1,1\}$. It is therefore convenient to define $S_1(t) := 2s_1(t) - 1$, so that $S_1(t) \in [-1,1]$. The condition '$s_1(t-1)$ is *larger than* $1/2$' then becomes '$S_1(t-1)$ is *larger than* 0'.

Modifying the If-part

The If-parts of the rules include four variables: $s_1(t-1)$, $s_4(t-3)$, $s_4(t-2)$, and $s_4(t-1)$, so the FARB must include 2^4 rules. The If-part of the first rule includes only one variable $s_1(t-1)$. We transform this into an equivalent set of eight rules by adding all the possible assignment combinations for the other three variables, while keeping the same output in all these rules. The other two rules are handled similarly. For example, one added rule is:

---

[1] It is also possible to use conditions in the form $\sigma^{-1}(s_1(t)) = \alpha$ and $\sigma^{-1}(s_1(t)) = -\alpha$. However, using inequalities makes it easier to transform this FRB into a FARB.

62   Knowledge-Based Design of ANNs

R_9: If $S_1(t-1)$ is *smaller than* 0 and $s_4(t-3)$ equals 1
    and $s_4(t-2)$ equals 1 and $s_4(t-1)$ equals 1,
    Then $\sigma^{-1}(s_1(t)) \geq \alpha$.

This yields an FRB with sixteen rules.

Modifying the Then-part

The output of every rule in the 16-rule FRB is either

$$\sigma^{-1}(s_1(t)) \geq \alpha, \quad \text{or} \quad \sigma^{-1}(s_1(t)) \leq -\alpha.$$

To make it a FARB, we must change every output to

$$\sigma^{-1}(s_1(t)) = a_0 \pm a_1 \pm a_2 \pm a_3 \pm a_4,$$

with the actual $+$ or $-$ sign determined according to the If-part. For example, the output of R_9 above is modified from $\sigma^{-1}(s_1(t)) \geq \alpha$ to $\sigma^{-1}(s_1(t)) = a_0 - a_1 + a_2 + a_3 + a_4$. Of course, the meaning of the Then-part should not change, so the values of the a_is must be chosen such that $a_0 - a_1 + a_2 + a_3 + a_4 \geq \alpha$. In this way, the sixteen rules yield sixteen inequalities on the a_is. It is easy to verify that one possible solution, satisfying all these inequalities, is:

$$a_0 = a_2 = a_3 = a_4 = \alpha, \quad \text{and} \quad a_1 = 3\alpha.$$

With these modifications, the set of fuzzy rules is now a FARB. Inferring, with the terms {*larger than, smaller than*} modeled as in (2.7), with $\alpha = 2$, yields

$$\sigma^{-1}(s_1(t)) = 2a_1\sigma(2S_1(t-1)) + 2a_2\sigma(2s_4(t-3)) + 2a_3\sigma(2s_4(t-2))$$
$$+ 2a_4\sigma(2s_4(t-1)) + a_0 - \sum_{i=1}^{4} a_i$$
$$= 6\alpha\sigma(4s_1(t-1) - 2) + 2\alpha\sigma(2s_4(t-3)) + 2\alpha\sigma(2s_4(t-2))$$
$$+ 2\alpha\sigma(2s_4(t-1)) - 5\alpha.$$

Again, we approximate the functions on the right-hand side using linear functions. This yields

$$s_1(t) = \sigma\big(5.04\alpha s_1(t-1) + 2\alpha s_2(t-1) + 2\alpha s_3(t-1)$$
$$+ 0.76\alpha s_4(t-1) - 3.52\alpha\big). \tag{6.3}$$

Summarizing, stating the required functioning of each neuron s_i as a FARB yields (6.1), (6.2), and (6.3). Letting $s_0 \equiv 1$ (the bias neuron), these equations can be written in the form:

$$s_i(t) = \sigma\left(\sum_{j=0}^{4} w_{ij}s_j(t-1)\right), \qquad i = 1, 2, 3, \tag{6.4}$$

with

$$W = \begin{bmatrix} -3.52\alpha & 5.04\alpha & 2\alpha & 2\alpha & 0.76\alpha \\ -2.62 & 0 & 0 & 5.26 & 0 \\ 0 & 0 & 0 & 0 & 2 \end{bmatrix}, \qquad \alpha \gg 0. \tag{6.5}$$

Clearly, this is an RNN. Given an input string x with length l, the RNN output is $f_{out} := s_1(l+1)$. The string is accepted if $f_{out} \leq 0.5$, and rejected, otherwise. The network is initialized with $s_1(1) = s_2(1) = s_3(1) = 0$.

The design of this RNN, although systematic, included several simplifications and omissions of small error terms. Nevertheless, the following result establishes the correctness of the final RNN.

Proposition 6.1. *Consider the RNN defined by* (5.4), (6.4), *and* (6.5). *For any* $\alpha \geq 5.7$ *this RNN correctly classifies any given binary string according to* L_4.

PROOF. See Appendix A.

6.2 The Modular Approach

The modular approach is based on using the FARB–ANN equivalence to realize basic functional modules using ANNs. These modules are designed once and for all, and their correctness is verified. The modules are then used as basic building blocks in designing more complex ANNs. This approach is motivated by the method used by Siegelmann to analyze the computational power of ANNs [160].

In this section, we describe the design of six basic modules, and then use them to design four compound RNNs that solve language recognition problems. For each of the designed modules, we describe its input, output, and desired functioning. We then restate the desired functioning in terms of an FRB, and transform this into a suitable FARB. Since the IO mapping of the FARB is equivalent to that of an ANN, this immediately yields a realization of the module as an ANN.

Throughout, we use FARBs that are equivalent to RNNs with activation function σ_L (see (2.11)). Note that

$$\sigma_L(z) = z, \quad \text{for all } z \in [0,1]. \tag{6.6}$$

6.2.1 The Counter Module

Input: A vector $\mathbf{x} \in \{0,1\}^m$, and a sequence $\mathbf{I}(t) \in \{0,1\}^m$, $t = 1, 2, \dots$.
Output: $f(n(t))$, where $n(t) := |\{k \in [1{:}t] : \mathbf{I}(k) = \mathbf{x}\}|$, and f is some invertible function.
In other words, the module counts the number of occurrences, $n(t)$, of the binary vector $\mathbf{x}$ in the stream of inputs: $\mathbf{I}(1), \mathbf{I}(2), \dots, \mathbf{I}(t)$. It returns $f(n(t))$, and $n(t)$ itself can be retrieved by calculating $f^{-1}(f(n(t)))$.

We use the specific function $f(z) := 2^{-z}$. To realize this module using an RNN, we restate the required functioning as a FARB with two output functions:

the first, $s_1(t)$, indicates whether $\mathbf{I}(t)$ equals $\mathbf{x}$ or not. The second, $s_2(t)$, is initialized to 1, and multiplied by $1/2$ at each time τ such that $\mathbf{I}(\tau) = \mathbf{x}$.

We design an FRB for each of these outputs. For $s_1(t)$, the rules are:

$$R_1 : \text{ If } I_1(t) \text{ equals } x_1 \text{ and } \ldots \text{ and } I_m(t) \text{ equals } x_m,$$
$$\text{Then } s_1(t+1) = 1, \qquad (6.7)$$
$$R_2 : \text{ Else, } s_1(t+1) = 0.$$

We modify this into a FARB as follows. First, we state 2^m rules covering, in their If-part, all the possible combinations of '$I_j(t)$ equals x_j' and '$I_j(t)$ not equals x_j'. We change the Then-part to $\sigma_L^{-1}(s_1(t+1)) = a_0 \pm a_1 \pm \ldots \pm a_m$, with the actual signs determined by the If-part. This yields the FARB:

$$R_1: \text{ If } I_1(t) \text{ equals } x_1 \text{ and } \ldots \text{ and } I_m(t) \text{ equals } x_m,$$
$$\text{Then } \sigma_L^{-1}(s_1(t+1)) = a_0 + a_1 + \ldots + a_m,$$
$$\vdots$$
$$R_{2^m}: \text{ If } I_1(t) \text{ not equals } x_1 \text{ and } \ldots \text{ and } I_m(t) \text{ not equals } x_m,$$
$$\text{Then } \sigma_L^{-1}(s_1(t+1)) = a_0 - a_1 - \ldots - a_m.$$

The a_is are determined as follows. We require that the output of Rule R_1 is ≥ 1, and that the output of all the other rules is ≤ 0. This yields 2^m conditions. It is easy to verify that

$$a_0 = 1 - m, \quad a_1 = a_2 = \cdots = a_m = 1 \qquad (6.8)$$

is a valid solution.

For $u, v \in [0, 1]$, the terms 'u equals v' and 'u not equals v' are modeled using the MFs:

$$\mu_=(u, v) := \begin{cases} 1 - |u - v|, & \text{if } |u - v| \leq 1, \\ 0, & \text{if } |u - v| > 1, \end{cases}$$
$$\mu_{\neq}(u, v) := 1 - \mu_=(u, v). \qquad (6.9)$$

In particular, the terms $\{equals\ 1, equals\ 0\}$ are modeled by

$$\mu_=(u, 1) = u, \quad \text{and} \quad \mu_=(u, 0) = 1 - u, \qquad (6.10)$$

respectively.

If u and v are binary then

$$\beta(u, v) := \frac{\mu_=(u, v) - \mu_{\neq}(u, v)}{\mu_=(u, v) + \mu_{\neq}(u, v)}$$
$$= 2\mu_=(u, v) - 1$$
$$= 1 - 2|u - v|$$
$$= 1 - 2\sigma_L(u + v - 2uv).$$

Thus, inferring the FARB and using (6.8) yields

$$\sigma_L^{-1}(s_1(t+1)) = 1 - 2\sum_{j=1}^{m} \sigma_L(I_j(t) + x_j - 2I_j(t)x_j),$$

or

$$s_1(t+1) = \sigma_L(1 - 2\sum_{j=1}^{m} \sigma_L(I_j(t) + x_j - 2I_j(t)x_j)). \tag{6.11}$$

Clearly, this describes a second-order RNN with inputs $I_1(t), \ldots, I_m(t)$ and $x_1, \ldots, x_m$. It is easy to verify, using the definition of the function σ_L, that if $I_j(t) = x_j$, for all $j \in [1:m]$, then $s_1(t+1) = 1$, and that $s_1(t+1) = 0$, otherwise. In other words, $s_1(t)$ in the RNN functions exactly as described in (6.7).

The rules for the second output, $s_2(t)$, follow directly from the description of its required functioning:

R_1: If $s_1(t)$ equals 1, Then $s_2(t+1) = s_2(t)/2$,
R_2: If $s_1(t)$ equals 0, Then $s_2(t+1) = s_2(t)$,

with $s_2(1) = 1$. This is actually a FARB with $a_0(t) = 3s_2(t)/4$, $a_1(t) = -s_2(t)/4$, and

$$\begin{aligned}
\beta_{=1}(y) &= -1 + 2\mu_{=1}(y) \\
&= -1 + 2y \\
&= -1 + 2\sigma_l(y),
\end{aligned}$$

where the last equation holds for any $y \in [0,1]$. Inferring yields

$$s_2(t+1) = s_2(t) - s_1(t)s_2(t)/2.$$

Since $s_1(t) \in \{0,1\}$, this implies that $s_2(t) \in [0,1]$ for all $t \geq 1$, so using (6.6) yields

$$s_2(t+1) = \sigma_L(s_2(t) - s_1(t)s_2(t)/2), \qquad s_2(1) = 1. \tag{6.12}$$

Summarizing, the counter module is realized by (6.11) and (6.12). Note that these two equations describe a second-order RNN. It is easy to verify that $s_2(t+2) = 2^{-n(t)}$, where $n(t)$ is the number of occurrences of $\mathbf{x}$ in the set $\{\mathbf{I}(1), \mathbf{I}(2), \ldots, \mathbf{I}(t)\}$.

Two interesting particular cases are:

1. The case $m = 1$, i.e., the input is a scalar $I(t) \in \{0,1\}$. Then (6.11) becomes

$$s_1(t+1) = \sigma_L(1 - 2\sigma_L((1 - 2x)I(t) + x)). \tag{6.13}$$

The resulting RNN is depicted in Fig. 6.1.

2. The case $m = 0$, i.e., there are no inputs. In this case, if we define $s_1(t) \equiv 1$ then (6.12) yields

$$s_2(t+1) = \sigma_L(s_2(t)/2), \qquad s_2(1) = 1, \tag{6.14}$$

so clearly $s_2(t+1) = 2^{-t}$.

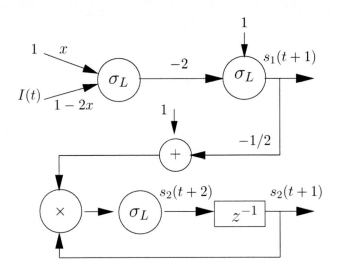

Fig. 6.1. Graphical representation of the counter module when $m = 1$

6.2.2 The Sequence-Counter Module

Input: a vector $\mathbf{x} \in \{0,1\}^m$, and a sequence $\mathbf{I}(t) \in \{0,1\}^m$, $t = 1, 2, \ldots$
Output: $f(l(t))$, where f is some invertible function and $l(t)$ is defined as follows.
If $\mathbf{I}(t) \neq \mathbf{x}$ then $l(t) = 0$. If $\mathbf{I}(t) = \mathbf{x}$ then $l(t) = l(t-1) + 1$.
In other words, $l(t)$ is the length of the *maximal* sequence $\{\mathbf{I}(t), \mathbf{I}(t-1), \ldots\}$, such that every element in this sequence equals $\mathbf{x}$.

To realize this using an FRB, we use two variables. The first, $s_1(t)$, indicates whether $\mathbf{I}(t) = \mathbf{x}$ or not, and so it is defined just as in the counter module above. The second variable, $s_2(t)$, is initialized to 1; divided by 2 at any time τ such that $\mathbf{I}(\tau) = \mathbf{x}$; and reset back to 1 if $\mathbf{I}(\tau) \neq \mathbf{x}$. Stating this as an FRB yields:

R_1: If $s_1(t)$ equals 1, Then $s_2(t+1) = s_2(t)/2$,
R_2: If $s_1(t)$ equals 0, Then $s_2(t+1) = 1$,

with initial values $s_1(1) = 0$, $s_2(1) = 1$.

This rule-base is a FARB. Inferring, using the MFs (6.10), and (6.6) yields

$$s_2(t+1) = \sigma_L(s_1(t)s_2(t)/2 - s_1(t) + 1), \qquad s_2(1) = 1. \qquad (6.15)$$

Summarizing, the sequence-counter is realized by (6.11) and (6.15). Note that these two equations describe a second-order RNN. It is easy to verify that indeed $s_2(t+2) = f(l(t))$, with $f(z) = 2^{-z}$.

6.2.3 The String-Comparator Module

Input: two binary sequences $I_1(t), I_2(t) \in \{0,1\}$, $t = 1, 2, \ldots$.
Output: at time $t+1$ the output is 1 if $I_1(\tau) = I_2(\tau)$ for all $\tau \in [1:t]$, and 0, otherwise.

In other words, this module indicates whether the two binary input sequences are identical or not.

We realize this module using a single variable $s_1(t)$. Stating the required functioning in terms of symbolic rules yields:

R_1: If $I_1(t)$ *equals* $I_2(t)$, Then $s_1(t+1) = s_1(t)$,
R_2: If $I_1(t)$ *not equals* $I_2(t)$, Then $s_1(t+1) = 0$,

with the initial value $s_1(1) = 1$.

This rule-base is a FARB. Inferring, using the MFs (6.9), and applying (6.6) yields

$$s_1(t+1) = \sigma_L(s_1(t)(1 - |I_1(t) - I_2(t)|)).$$

Since $I_1(t), I_2(t) \in \{0, 1\}$, this is equivalent to

$$s_1(t+1) = \sigma_L(s_1(t)s_2(t)), \qquad s_1(1) = 1, \tag{6.16}$$

where

$$s_2(t) := \sigma_L(1 - I_1(t) - I_2(t) + 2I_1(t)I_2(t)). \tag{6.17}$$

Summarizing, the string-comparator module is realized using the second-order RNN given by (6.16) and (6.17). It is easy to verify that if $I_1(\tau) = I_2(\tau)$, for all $\tau \in [1\!:\!t]$, then $s_1(t+1) = 1$, and that $s_1(t+1) = 0$, otherwise.

6.2.4 The String-to-Num Converter Module

Input: $I(t) \in \{0, 1\}$, $t = 1, 2, \dots$.

Output: the output at time t is $f(t) := \sum_{\tau=1}^{t} I(\tau)2^{-\tau}$.

In other words, this module calculates the value represented by the input binary string.

To realize this, we use an FRB with output $s_1(t)$, initialized to $s_1(t) = 0$. If $I(\tau) = 1$, then $2^{-\tau}$ is added to $s_1(t)$. If $I(\tau) = 0$, $s_1(t)$ is left unchanged. Stating this as an FRB yields:

R_1: If $I(t)$ *equals* 1, Then $s_1(t+1) = s_1(t) + s_2(t)/2$,
R_2: If $I(t)$ *equals* 0, Then $s_1(t+1) = s_1(t)$,

where $s_2(t)$ is defined by (6.14).

This rule-base is a FARB. Inferring, using the MFs (6.10), and applying (6.6) yields

$$s_1(t+1) = \sigma_L(s_1(t) + I(t)s_2(t)/2), \qquad s_1(1) = 0. \tag{6.18}$$

Summarizing, this module is realized using the second-order RNN given by (6.14) and (6.18). It is easy to verify that indeed $s_1(t+1) = \sum_{\tau=1}^{t} I(\tau)2^{-\tau}$.

68   Knowledge-Based Design of ANNs

6.2.5   The Num-to-String Converter Module

Input: a value $x \in [0, 1]$.

Output: at time t the output is $y(t) \in \{0, 1\}$ such that $0.y(1)y(2) \ldots$ is the binary representation of x. In other words, this module is just the opposite of the String-to-Num module.

We realize this module using two FRBs. The first FRB has output $s_1(t)$, initialized to $s_1(1) = x$. At each time t, the value of $s_1(t)$ indicates whether the next digit, i.e., $y(t+1)$ should be zero or one. The second FRB, with output $s_2(t)$, keeps track of the digit itself and shifts the value stored in $s_1(t)$ one bit to the left, in order to expose the next digit.

Stating the required functioning of these two variables as FRBs yields:

$R_1^{s_2}$: If $s_1(t)$ is *larger than* $1/2$, Then $s_2(t+1) = 1$,
$R_2^{s_2}$: If $s_1(t)$ is *smaller than* $1/2$, Then $s_2(t+1) = 0$,

and

$R_1^{s_1}$: If $s_2(t)$ *equals* 1, Then $s_1(t+1) = 2(s_1(t) - 1/2)$,
$R_2^{s_1}$: If $s_2(t)$ *equals* 0, Then $s_1(t+1) = 2s_1(t)$,

with $s_1(1) = x$ and $s_2(1) = 0$.

Both these rule-bases are FARBs. For the first FARB, we use the MFs (2.12), with $\Delta = 1/2$ and $k = 1$, for the terms $\{larger\ than\ 1/2,\ smaller\ than\ 1/2\}$. Then (2.13) yields $\beta = 2\sigma_L(y - 1/2) - 1$. For the second FARB, we use the MFs (6.10) for the terms $\{equals\ 1,\ equals\ 0\}$. Inferring and using (6.6) then yields

$$s_1(t+1) = \sigma_L(2s_1(t) - s_2(t)), \quad s_1(1) = x,$$
$$s_2(t+1) = \sigma_L(s_1(t) - 1/2), \quad s_2(1) = 0. \tag{6.19}$$

The binary representation itself is obtained by:

$$y(t) = \begin{cases} 1, & \text{if } s_2(t) \geq 0, \\ 0, & \text{if } s_2(t) < 0. \end{cases} \tag{6.20}$$

It is easy to verify that the RNN given by (6.19) and (6.20) indeed realizes the Num-to-String module.

All the modules designed above were defined in a crisp manner. Stating the required functioning in terms of a FARB can also lead, quite naturally, to modules with a fuzzy mode of operation. This is demonstrated in the next module.

6.2.6   The Soft Threshold Module

Input: a value $b \in [0, 1]$ and a sequence $I(t) \in [0, 1]$, $t = 1, 2, \ldots$

Output: the output at time $t + 1$ is 1 if $I(\tau)$ is *much larger* than b for all $\tau \in [1{:}t]$. Otherwise, the output is 0.

Note the deliberate use of the verbal term *much larger* here. We realize this using an FRB with output $s_1(t)$, initialized to $s_1(1) = 1$. If $I(t)$ is *much larger* than b, then $s_1(t+1)$ is left unchanged. If $I(t)$ is *not much larger* than b, then $s_1(t+1)$ is set to 0. We state this as:

R_1: If $(I(t) - b)$ is *positive*, Then $s_1(t+1) = s_1(t)$,
R_2: If $(I(t) - b)$ is *negative*, Then $s_1(t+1) = 0$.

This is a FARB. Inferring, using the MFs (2.10) for the terms {*positive, negative*}, and applying (6.6) yields

$$s_1(t+1) = \sigma_L(s_1(t)s_2(t)), \qquad s_1(1) = 1, \qquad (6.21)$$

with

$$s_2(t) := \sigma_L\left(\frac{1}{2} + \frac{I(t) - b}{2\Delta}\right). \qquad (6.22)$$

Summarizing, the soft threshold module is realized by the second-order RNN given by (6.21) and (6.22). If $I(t) \gg b$, $s_2(t) \approx 1$ and then $s_1(t+1) \approx s_1(t)$. Thus, if $I(t) \gg b$ for all t, then $s_1(t+1) \approx s_1(1) = 1$. On the other-hand, if $I(t) \ll b$, $s_2(t) \approx 0$ and then $s_1(t+1) \approx 0$. This agrees with the desired functioning of the soft-threshold module.

Note that

$$\lim_{\Delta \to 0} s_2(t) = \begin{cases} 0, & \text{if } I(t) < b, \\ 1/2, & \text{if } I(t) = b, \\ 1, & \text{if } I(t) > b, \end{cases}$$

and then (6.21) implies that the module becomes a hard threshold unit.

The modular approach to KBD of RNNs is based on using the basic modules as building blocks. Since each module is realized as an RNN, the result is a hierarchal network that constitutes several simple RNNs. The following examples demonstrate how our set of modules can be used to design RNNs that solve nontrivial problems.

6.2.7 KBD of an RNN for Recognizing the Extended L_4 Language

Fix an integer $n > 0$. The *extended L_4 language* is the set of all binary strings that do *not* contain a substring of n consecutive zeros (for the particular case $n = 3$, this is just the L_4 language). We apply the modular approach to design an RNN that recognizes this language.

Detecting that an input string $I(1), I(2), \ldots$ contains a substring of n consecutive zeros can be easily done using a Sequence-Counter Module followed by a Soft Threshold Module (see Fig. 6.2). In the Sequence-Counter Module, $x = 0$, so that its output is $2^{-l(t)}$, where $l(t)$ is the length of the maximal subsequence of consecutive zeros in the input sequence.

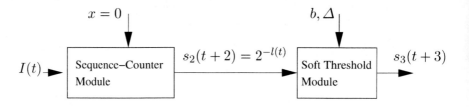

Fig. 6.2. Modular RNN for recognizing the extended L_4 language

Expressing the modules in terms of the RNNs that realize them (see (6.13), (6.15), (6.21), and (6.22)) yields the complete network:

$$s_1(t+1) = \sigma_L(1 - 2\sigma_L(I(t))),$$
$$s_2(t+1) = \sigma_L(s_1(t)s_2(t)/2 - s_1(t) + 1),$$
$$s_3(t+1) = \sigma_L(s_3(t)s_4(t)),$$
$$s_4(t) = \sigma_L\left(\frac{1}{2} + \frac{s_2(t) - b}{2\Delta}\right), \quad (6.23)$$

with $s_2(1) = s_3(1) = 1$. We set $b = (2^{-n} + 2^{-(n-1)})/2$. Then $s_2(t+2) - b = 2^{-l(t)} - b$ changes sign exactly when $l(t) = n$. Furthermore, since $s_2(t+2) \in \{2^0, 2^{-1}, 2^{-2}, \dots\}$, we can find $\Delta > 0$ small enough such that $\sigma_L(1/2 + (s_2(t) - b)/2\Delta) \in \{0, 1\}$, and then s_3 and s_4 actually realize a hard threshold unit.

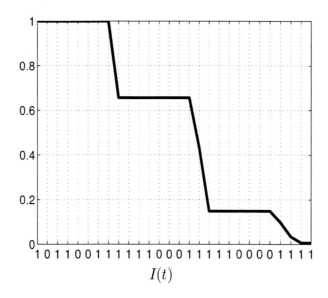

Fig. 6.3. Output $s_3(t)$ of RNN (6.23) for the input sequence shown on the x axis

Finally, the classification rule is: Accept a given string of length k, if $s_3(k+3) = 1$,[2] and Reject, otherwise. The correctness of this RNN follows immediately from the correctness of the constituting modules.

Note that increasing the value of Δ (i.e., "softening" the threshold module) turns (6.23) into a *fuzzy recognizer*. Strings with sequences of zeros much longer than n will be rejected, and strings with sequences of zeros much shorter than n will be accepted. For the intermediate cases, the output is $s_3(k+3) \in (0,1)$, with the exact value depending on the number and length of the zero sequences in the input.

To demonstrate this, Fig. 6.3 depicts the output of the RNN (6.23) with $n = 3, b = (2^{-n} + 2^{-(n-1)})/2 = 3/16$, and $\Delta = 0.2$, for the input shown on the x axis. It may be seen that $s_3(k+3)$ decreases whenever a sequence of two or more consecutive zeros appears in the output. Longer sequences yield a greater decrease in $s_3(k+3)$.

6.2.8 KBD of an RNN for Recognizing the AB Language

The AB language is the set of all binary strings that include an equal number of zeros and ones (e.g., "010110"). We apply the modular approach to design an RNN that recognizes this language. Note that the AB language is context-free, and *not* a regular language, so common KBD techniques that are based on the DFA-to-RNN conversion [2; 127] cannot be applied at all.

We realize a suitable RNN by feeding the input string $I(1), I(2), \ldots$ into two counter modules. The first with $x = 0$, so that its output is $2^{-n_0(t)}$, where n_0 denotes the number of zeros in the input sequence. The second counter has $x = 1$. The string is accepted iff the counter outputs are equal (see Fig. 6.4). Expressing each module in terms of the RNN that realizes it yields the complete network:

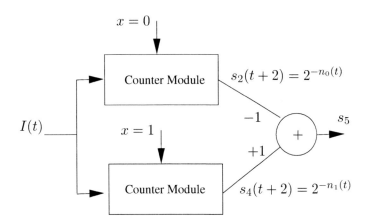

Fig. 6.4. Modular RNN for recognizing the AB language

---
[2] The shift to time $k+3$ is required to accommodate the processing time of the RNN.

$$s_1(t+1) = \sigma_L(1 - 2\sigma_L(I(t))),$$
$$s_2(t+1) = \sigma_L(s_2(t) - s_1(t)s_2(t)/2),$$
$$s_3(t+1) = \sigma_L(1 - 2\sigma_L(1 - I(t))), \qquad (6.24)$$
$$s_4(t+1) = \sigma_L(s_4(t) - s_3(t)s_4(t)/2),$$
$$s_5(t+1) = s_4(t) - s_2(t),$$

with $s_i(1) = 1$, $i = [1:4]$.

The classification rule for a string of length k is: Accept if $s_5(k+3) = 0$, and Reject, otherwise. The correctness of (6.24) follows immediately from the correctness of the constituting modules.

6.2.9 KBD of an RNN for Recognizing the Balanced Parentheses Language

Let $n_0(t)$ ($n_1(t)$) denote the number of zeros (ones) in the input string $\{I(0), \ldots, I(t)\}$. The *balanced parentheses language* (BPL) includes all the strings that satisfy: $n_0(t) = n_1(t)$, and $n_0(\tau) \geq n_1(\tau)$ for all $\tau \leq t$. If the symbol 0 (1) represents left (right) parenthesis, then a string is in the BPL if each right parenthesis has a corresponding left parenthesis that occurs *before* it.

We use the modular approach to design an RNN for recognizing the BPL. Note that the BPL is a *non-regular*, context-free, language, so common KBD-to-RNN techniques cannot be applied for KBD of an RNN that recognizes this language.

The RNN contains two counters for calculating $2^{-n_0(t)}$ and $2^{-n_1(t)}$. The difference $2^{-n_1(t)} - 2^{-n_0(t)}$ is fed into a soft threshold module (see Fig. 6.5). The values b and Δ are set so that the output of this module will be 1 only if

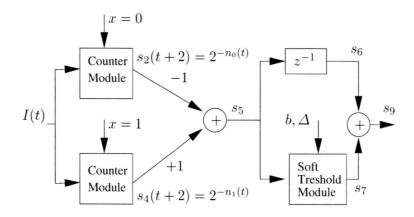

Fig. 6.5. Modular RNN for recognizing the balanced parentheses language

$$n_0(\tau) \geq n_1(\tau), \qquad \text{for all } \tau \leq t,$$

or, equivalently,

$$2^{-n_0(\tau)} \leq 2^{-n_1(\tau)}, \qquad \text{for all } \tau \leq t.$$

Expressing each basic module in the terms of the RNN that realizes it yields the complete network:

$$
\begin{aligned}
s_1(t+1) &= \sigma_L(1 - 2\sigma_L(I(t))), \\
s_2(t+1) &= \sigma_L(s_2(t) - s_1(t)s_2(t)/2), \\
s_3(t+1) &= \sigma_L(1 - 2\sigma_L(1 - I(t))), \\
s_4(t+1) &= \sigma_L(s_4(t) - s_3(t)s_4(t)/2), \\
s_5(t+1) &= s_4(t) - s_2(t), \\
s_6(t+1) &= s_5(t), \\
s_7(t+1) &= \sigma_L(s_7(t)s_8(t)), \\
s_8(t) &= \sigma_L\left(\frac{1}{2} + \frac{s_5(t) - b}{2\Delta}\right), \\
s_9(t+1) &= s_7(t) + s_6(t),
\end{aligned}
\tag{6.25}
$$

with $s_i(1) = 1$, $i \in [1:9]$. Here the neuron s_6 realizes a pure delay which is needed to synchronize the output streams of the counters and of the soft threshold module.

The classification rule for a string of length k is: Accept if $s_9(k+5) = 1$, and Reject, otherwise. The following result analyzes the correctness of the RNN.

Proposition 6.2. *Fix an integer $l > 0$. Suppose that the parameters b and Δ of the soft threshold module satisfy*

$$\Delta > 0, \qquad \Delta - 2^{-l} \leq b \leq -\Delta. \tag{6.26}$$

Then the RNN (6.25) correctly classifies all strings with length $\leq l$ according to the BPL.

PROOF. Consider a string with length l. It follows from (6.25) that

$$
\begin{aligned}
s_9(l+5) &= s_7(l+4) + s_6(l+4) \\
&= s_7(l+4) + s_5(l+3) \\
&= s_7(l+4) + s_4(l+2) - s_2(l+2) \\
&= \sigma_L(s_7(l+3)s_8(l+3)) + 2^{-n_1(l)} - 2^{-n_0(l)}.
\end{aligned}
\tag{6.27}
$$

We consider two cases.

Case 1. The string belongs to the BPL. In this case, $n_0(l) = n_1(l)$ and $2^{-n_0(\tau)} \leq 2^{-n_1(\tau)}$ for all $\tau \in [1, l]$. It follows that $s_5(\tau) = s_4(\tau - 1) - s_2(\tau - 1) \geq 0$, so (6.26) yields

$$\frac{s_5(\tau) - b}{2\Delta} \geq \frac{-b}{2\Delta} \geq 1/2, \qquad \text{for all } \tau.$$

74   Knowledge-Based Design of ANNs

Hence, $s_8(\tau) = 1$ and $s_7(\tau) = 1$ for all $\tau \in [1, l]$, and (6.27) yields

$$s_9(l + 5) = 1,$$

i.e., the RNN classification is correct.

Case 2. The string does not belong to the BPL. That is, either $n_0(\tau) < n_1(\tau)$ for some $\tau \in [1, l]$, or $n_0(l) \neq n_1(l)$. We will show that in this case $s_9(l+5) \neq 1$. We consider two sub-cases.

Case 2.1. Suppose that $n_0(\tau) < n_1(\tau)$ for some $\tau \in [1, l]$. Then, $n_0(\tau) \leq n_1(\tau) - 1$, so $2^{-n_1(\tau)} - 2^{-n_0(\tau)} \leq -2^{-n_1(\tau)}$. This implies that

$$2^{-n_1(\tau)} - 2^{-n_0(\tau)} \leq -2^{-\tau} \leq -2^{-l}. \tag{6.28}$$

Substituting (6.28) in (6.25) yields

$$s_8(\tau + 3) = \sigma_L \left(\frac{1}{2} + \frac{2^{-n_1(\tau)} - 2^{-n_0(\tau)} - b}{2\Delta} \right)$$

$$\leq \sigma_L \left(\frac{1}{2} + \frac{-2^{-l} - b}{2\Delta} \right).$$

It follows from (6.26) that $s_8(\tau + 3) = 0$. This implies that $s_7(t) = 0$ for all $t \geq \tau + 3$, so

$$s_9(l + 5) = 2^{-n_1(l)} - 2^{-n_0(l)} \neq 1.$$

Case 2.2. Suppose that $n_0(l) \neq n_1(l)$. The analysis above shows that $\sigma_L(s_7(l + 3)s_8(l + 3)) \in \{0, 1\}$. Thus, either $s_9(l + 5) = 2^{-n_1(l)} - 2^{-n_0(l)}$ or $s_9(l + 5) = 1 + 2^{-n_1(l)} - 2^{-n_0(l)}$. This implies that when $n_0(l) \neq n_1(l)$, $s_9(l + 5) \neq 1$.

This completes the proof for the case where the length of the string is l. It is straightforward to see that the same result holds when the length is $\leq l$. □

6.2.10   KBD of an RNN for Recognizing the $0^n 1^n$ Language

The $0^n 1^n$ language is the set of all binary strings of the form

$$\underbrace{00 \ldots 0}_{n \text{ times}} \underbrace{11 \ldots 1}_{n \text{ times}}$$

for some n. The $0^n 1^n$ language is context-free.

We now apply the modular approach to design an RNN that recognizes this language. Let $n_0(t)$ ($n_1(t)$) denote the number of 0s (1s) in the input string $\{I(1), \ldots, I(t)\}$, and let $l_1(t)$ denote the length of the *maximal* sequence $\{I(t), I(t-1), \ldots\}$, such that every element in this sequence equals 1. A string $\{I(1), \ldots, I(t)\}$ is in the $0^n 1^n$ language iff it satisfies two conditions:

$$n_0(t) = n_1(t),$$
$$l_1(t) = n_1(t).$$

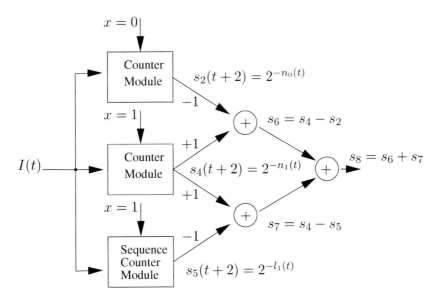

Fig. 6.6. Modular RNN for recognizing the $0^n 1^n$ language

It follows that we can realize a suitable RNN by feeding the input string into two counters and one sequence-counter. The counters output n_0 and n_1. The sequence counter outputs l_1. The input string belongs to the $0^n 1^n$ language iff the outputs of the counters are equal, and the outputs of the counter calculating n_1 equals the output of the sequence counter. The resulting RNN is depicted in Fig. 6.6.

Expressing each module in terms of the RNN that realizes it yields the complete network:

$$\begin{aligned}
s_1(t+1) &= \sigma_L(1 - 2\sigma_L(I(t))), \\
s_2(t+1) &= \sigma_L(s_2(t) - s_1(t)s_2(t)/2), \\
s_3(t+1) &= \sigma_L(1 - 2\sigma_L(1 - I(t))), \\
s_4(t+1) &= \sigma_L(s_4(t) - s_3(t)s_4(t)/2), \\
s_5(t+1) &= \sigma_L(s_3(t)s_5(t)/2 - s_3(t) + 1), \\
s_6(t+1) &= s_4(t) - s_2(t), \\
s_7(t+1) &= s_4(t) - s_5(t), \\
s_8(t+1) &= s_6(t) + s_7(t).
\end{aligned} \quad (6.29)$$

The classification rule for a string of length k is: Accept if $s_8(k+4) = 0$, and Reject, otherwise. The next result shows that the RNN indeed functions properly.

Proposition 6.3. *The RNN (6.29) correctly classifies any string according to the $0^n 1^n$ language.*

76   Knowledge-Based Design of ANNs

PROOF. Let t denote the length of the input string. It follows from (6.29) and the correctness of the constituting modules that

$$s_8(t+4) = s_6(t+3) + s_7(t+3)$$
$$= 2s_4(t+2) - s_2(t+2) - s_5(t+2)$$
$$= 2 \cdot 2^{-n_1(t)} - 2^{-n_0(t)} - 2^{-l_1(t)}. \qquad (6.30)$$

We consider two cases.

CASE 1. Suppose that the input string belongs to the 0^n1^n language. In other words, $n_0(t) = n_1(t) = l_1(t)$. Eq. (6.30) implies that $s_8(t+4) = 0$, i.e. the RNN classification is correct.

CASE 2. Suppose that the input string does not belong to the 0^n1^n language. We need to show that $s_8(t+4) \neq 0$. Seeking a contradiction, assume that $s_8(t+4) = 0$. Then (6.30) yields

$$2 - 2^{n_1(t)-n_0(t)} = 2^{n_1(t)-l_1(t)}. \qquad (6.31)$$

If $n_1(t) = l_1(t)$ then (6.31) yields $n_1(t) = n_0(t)$, but this contradicts the fact that the string does not belong to the 0^n1^n language. If $n_1(t) \neq l_1(t)$, then by the definition of $n_1(t)$ and $l_1(t)$, $n_1(t) > l_1(t)$. Now (6.31) implies that $2^{n_1(t)-n_0(t)} < 0$, which is a contradiction. $\qquad \square$

7 Conclusions and Future Research

The ability of ANNs to learn and generalize from examples, and to generate robust solutions, makes them very suitable in a diversity of applications where algorithmic approaches are either unknown or difficult to implement. A major drawback, however, is that the knowledge learned by the network is represented in an exceedingly opaque form, namely, as a list of numerical coefficients. This black-box character of ANNs hinders the possibility of their more wide-spread acceptance. The problem of extracting the knowledge embedded in the ANN in a comprehensible form has been intensively addressed in the literature.

Another drawback of ANNs is that standard training algorithms do not guarantee convergence, and are highly dependent on the initial values of the networks parameters. Efficient methods for determining the initial architecture and parameter values of ANNs are quite important, as they may improve the trained ANN generalization capability, and reduce training times. In many problems some initial information is known, and an important problem is how this knowledge can be used in order to design an initial ANN.

In this work, we introduced a novel fuzzy rule-base, the FARB, and showed that its IO mapping is mathematically equivalent to that of an ANN. We used this equivalence to develop new approaches for: (1) extracting knowledge from trained ANNs, and representing it in a comprehensible form; and (2) knowledge-based design of ANNs based on prior knowledge. These applications were illustrated for both feedforward ANNs and first- and second-order recurrent ANNs.

For large-scale networks, the corresponding FARB may include either a large number of rules or complicated rules, and thus hamper the FARB comprehensibility. In order to minimize the effect of this manifestation of the *curse of dimensionality* [11], we also presented a systematic procedure for rule reduction and simplification.

7.1 Future Research

The FARB is based on standard tools from the field of fuzzy logic, and the FARB–ANN equivalence holds for a large variety of standard ANNs. This opens

78 Conclusions and Future Research

the door to many more potential applications which are based on transferring concepts and ideas from fuzzy logic theory to ANNs, and vice versa.

7.1.1 Regularization of Network Training

The ANN–FARB equivalence provides a convenient tool for extracting symbolic knowledge from trained ANNs. It is highly desirable that the extracted rule-base include a small number of rules that are as simple as possible. Two potential approaches for achieving this goal are: (1) first extract the FARB and then apply some simplification procedure (as described in Chapter 4); and (2) during training, add regularization terms [150] that will force the ANN to develop a skeletal form, and then extract the equivalent FARB.

When using the latter approach, it is of course possible to use regularization terms that are standard in the ANN literature, e.g., $\sum_{i,j} w_{ij}^2$ [69]. However, if we know beforehand that the KE approach is based on the ANN–FARB equivalence, then it is natural to try and develop specialized regularization terms that are particularly suitable for this approach, that is, terms that will potentially yield a simple as possible FARB.

To make this more concrete, consider a FARB with the fuzzy terms *larger than* 0 and *smaller than* 0 (we assume that the bias terms are included in the x_is). A typical rule is then:

R_k: If x_1 is *larger than* 0 and x_2 is *larger than* 0 and ... and x_m is *larger than* 0, Then $f = \ldots$

Suppose that there exist two indexes $i \neq j$ and a number $\alpha > 0$ such that

$$x_i \cong \alpha x_j, \tag{7.1}$$

for all the training samples.

Let R_k' denote the rule that is obtained from R_k by deleting the the jth atom. Condition (7.1) then implies that R_k and R_k' will have similar DOFs, since a high (low) truth value for the atom 'x_i is *larger than* 0' will correspond to a high (low) truth value for the atom 'x_i is *smaller than* 0'. Thus, the jth atom can be deleted from the FARB, yielding a simpler FARB, with little effect on the output.

We would like to add a regularization term that will force the ANN to satisfy (7.1). For two vectors $\mathbf{g}, \mathbf{h} \in \mathbb{R}^n \setminus \{\mathbf{0}\}$, let

$$b(\mathbf{g}, \mathbf{h}) := \frac{\mathbf{g}^T \mathbf{h}}{||\mathbf{g}|| \, ||\mathbf{h}||},$$

that is, $b = \cos(\theta)$, where θ is the angle between $\mathbf{g}$ and $\mathbf{h}$. Recalling that $x_i = (\mathbf{w}^i)^T \mathbf{z}$, where $\mathbf{w}^i = (w_{i1}, \ldots, w_{in})^T$, and $\mathbf{z}$ is the ANN input, we see that requiring (7.1) is identical to requiring that $b(\mathbf{w}^i, \mathbf{w}^j)$ is maximized.

This suggests that a suitable cost criterion might be

$$E - \lambda \sum_{i,j} b(\mathbf{w}^i, \mathbf{w}^j),$$

where E is a standard error function (e.g., the squared difference between the desired and actual output), and $\lambda > 0$ is a weighting factor. It may be interesting to study the effect of using this, or other specially designed regularization terms, on the comprehensibility of the extracted FARB.

7.1.2 Extracting Knowledge during the Learning Process

Standard learning algorithms for ANNs have several drawbacks. Convergence is not guaranteed, and even if it occurs, the stopping point might be a local minimum in the weights space. Furthermore, it is not clear when to terminate the learning process and how to assess the generalization capability of the network.

ANN training is a dynamic process and there is considerable interest in understanding how it evolves in time, and how different parameter values (e.g., the momentum factor) affect it. For example, Pollack [135] has studied training an ANN to recognize binary strings with an odd number of 1s, and showed that the trained ANN goes through a type of *phase-transition*. At a certain point during the training, there is a sharp transition from parameter values that yield poor performance to values that yield good generalization.

Since the equivalence between the ANN and a suitable FARB holds at any moment, it is possible to express the knowledge embedded in the ANN after *each iteration of the learning algorithm*. In this way, the time evolution of the ANN during training becomes a time evolution of a *symbolic rule-base*. Analyzing this dynamic symbolic representation of the knowledge embedded in the ANN may provide a deeper understanding of the learning process.

7.1.3 Knowledge Extraction from Support Vector Machines

One of the most successful descendants of ANNs are *support vector machines* (SVMs) [22; 158]. SVMs have been widely applied for pattern recognition, regression, and classification. SVMs can be trained using a set of classified examples $\{\mathbf{x}^i, y_i\}_{i=1}^n$, where y_i is the desired output when the input is $\mathbf{x}^i$. The input vectors are mapped into a higher dimensional space, and are then classified using a maximal separating hyperplane. After training, the SVM output is a function of a *subset* of the samples, called the *support vectors*. SVMs have demonstrated an impressive success in various classification problems, but their comprehensibility is quite poor. This raises the problem of KE from trained SVMS (see, e.g., [9; 123]).

An SVM realizes a mapping in the form:

$$f(\mathbf{x}) = b + \sum_{\mathbf{x}^i \in S} \alpha_i y_i K\left(\mathbf{x}, \mathbf{x}^i\right), \tag{7.2}$$

where S is the set of support vectors, $K(\cdot, \cdot)$ is a *kernel function*, and $\alpha_i, b \in \mathbb{R}$. Commonly used kernel functions include radial basis functions $K(\mathbf{x}, \mathbf{x}^i) = \exp(-\gamma \left\| \mathbf{x} - \mathbf{x}^i \right\|^2)$, and sigmoid functions $K(\mathbf{x}, \mathbf{x}^i) = \tanh(\kappa \mathbf{x}^T \mathbf{x}^i + c)$.

80   Conclusions and Future Research

Comparing (7.2) with (3.1) reveals a similarity between the IO mapping of SVMs and feedforward ANNs. Hence, techniques that extract knowledge from feedforward ANNs may potentially be adapted for KE from SVMs [23]. It would be interesting to explore possible connections between SVMs and FARBs.

7.1.4 Knowledge Extraction from Trained Networks

Neural networks have been successfully applied in numerous applications (see, e.g., [99; 120]). In many cases it would be beneficial to extract the knowledge from the trained networks and represent it in a comprehensible form. Specific examples include the following.

- *Biological and medical systems.* Neural modeling of biological and medical systems is a very active research area (see, e.g., [137; 63]). However, a major obstacle to medical applications is the black-box character of ANNs. Obtaining more knowledge on the internal features and the IO mapping of the network is highly desirable. Extracting the knowledge embedded in an ANN that models a biological system may lead to a deeper understanding of this system. Interesting examples include ANNs that model the effect of various diseases by inducing specific artificial "damage" to the ANN. Using the ANN–FARB equivalence makes it possible to extract symbolic information from both the original and damaged ANNs. Comparing this information may lead to a more comprehensible description of what exactly the damage is, and thus improve the understanding of the actual effect caused by the disease.

  More specifically, Alzheimer's disease, the most common form of dementia, is intensively investigated and modeled using various techniques. Two neural models are the *synaptic deletion and compensation* [64; 146], and the *synaptic runaway* [57; 58], both with interesting results and comparisons to clinical observations. These models were even used to derive therapeutic suggestions [34] (e.g., minimize new memory load). Exploring such networks using the FARB–ANN equivalence may offer new directions for further research.
- *High-level image processing.* Neural networks are used for facial detection [142], land cover classification [82], and more [99]. It is interesting to understand which visual features the ANN learned to detect and use. The FARB formulation may be useful for addressing this issue.
- *Speech and language modeling.* The recognition and understanding of human speech and natural language is a necessary step toward many AI systems and human-machine interfaces. Neural networks have been used for speaker recognition and classification [48; 76], text-to-speech and speech-to-text transformations [154], and more [99]. Again, understanding the network *modus operandi* might help to improve the system robustness, performance, and widespread use.
- *Stochastic networks.* In stochastic ANNs, the output is a random variable. Its value is drawn from some probability function parametrized by the neurons inputs [182]. Logistic functions are commonly used as probability mass

functions, so the network can be modeled using a FARB, whose output is the *probability function*. Thus, the FARB formulation may be a useful tool for KBN in stochastic nets.

- *Knowledge refinement*. In many practical engineering tasks, prior knowledge about the problem is used. However, this knowledge is often vague, partial, and may contain incorrect information. Improving this knowledge may lead to better solutions, faster convergence, etc. If a classified set of samples concerning the problem is available, *knowledge refinement* can be used to modify the prior knowledge by making it more consistent with the given examples. The FARB–ANN equivalence may also be used for knowledge refinement via the following procedure: (1) state the initial knowledge as a FARB; (2) inference to obtain the corresponding ANN; (3) train this ANN using given data; and (4) use the FARB–ANN equivalence again to extract the refined knowledge in symbolic form. Note that FRBs may be particularly useful for expressing initial knowledge due to their inherent ability to handle vagueness and uncertainty [122].

- *Automating the KE process*. The conversion of an ANN into the corresponding FARB and the FARB simplification procedure are well-defined and structured processes. An interesting research direction is to code these processes as computer procedures and thus to develop a fully automatic approach for extracting comprehensible information from ANNs.

- *KBN of ANNs with several hidden layers*. In this work, we considered using the FARB–ANN equivalence for KBN in ANNs with a single hidden layer. Yet, the approach can be extended to handle ANNs with several hidden layers. Indeed, we can represent the output f_i of each neuron in the first hidden layer using an appropriate FARB, say, $FARB_i$. The output t_i of each neuron in the second hidden layer can now be described as a FARB with inputs f_i, and so on. Thus, an ANN with multiple hidden layers is equivalent to a *hierarchy* of FARBs.

A Proofs

Proof of Proposition 5.6

We require the following result.

Lemma A.1. *Consider the RNN defined by* (5.3) *and* (5.4) *with* $s_i(1) = 0$, $i \in [1 : 3]$. *Suppose that there exist* $\epsilon_1, \epsilon_2 \in [0, 1/2)$ *such that the following conditions hold:*

1. *If* $s_1(t - 1) \geq 1 - \epsilon_1$, *then* $s_1(t) \geq 1 - \epsilon_1$,
2. *If* $s_4(t - 1) = s_4(t - 2) = s_4(t - 3) = 1$, *then* $s_1(t) \geq 1 - \epsilon_1$,
3. *If Conditions 1 and 2 do not hold, and* $s_1(t - 1) \leq \epsilon_2$, *then* $s_1(t) \leq \epsilon_2$.

Then, the RNN correctly classifies any given binary string according to the L_4 *language.*

PROOF. Consider an arbitrary input string. Denote its length by l. We consider two cases.

Case 1: The string does *not* include '000' as a substring. In this case, the If-part in Condition 2 is never satisfied. Since $s_1(1) = 0$, Condition 3 implies that $s_1(t) \leq \epsilon_2$, for $t = 1, 2, 3 \ldots$, hence, $s_1(l + 1) \leq \epsilon_2$. Recalling that the network output is $f_{out} = s_1(l + 1)$, yields $f_{out} \leq \epsilon_2$.

Case 2: The string contains a '000' substring, say, $I(m - 2)I(m - 1)I(m) =$'000', for some $m \leq l$. Then, according to Condition 2, $s_1(m + 1) \geq 1 - \epsilon_1$. Condition 1 implies that $s_1(t) \geq 1 - \epsilon_1$ for $t = m + 1, m + 2, \ldots$, so $f_{out} \geq 1 - \epsilon_1$.

Summarizing, if the input string includes a '000' substring, then $f_{out} \geq 1 - \epsilon_1 > 1/2$, otherwise, $f_{out} \leq \epsilon_2 < 1/2$, so the RNN accepts (rejects) all the strings that do (not) belong to the language. □

We now prove Proposition 5.6 by showing that the RNN defined by (5.3), (5.4), and (5.5) indeed satisfies the three conditions in Lemma A.1. Note that using (5.5) yields

$$s_1(t) = \sigma \left(15.2 s_1(t - 1) + 8.4 s_2(t - 1) + 0.2 s_3(t - 1) + 3 s_4(t - 1) - 7.6 \right), \tag{A.1}$$

E. Kolman, M. Margaliot: Knowledge-Based Neurocomputing, STUDFUZZ 234, pp. 83–86.
springerlink.com © Springer-Verlag Berlin Heidelberg 2009

84 Proofs

whereas substituting (5.4) in (5.6) and (5.7) yields

$$s_4(t) \in \{-1, 1\}, \ s_3(t) \in \{0.015, 0.98\}, \quad \text{and} \quad s_2(t) \in [0, 0.8]. \tag{A.2}$$

Suppose that

$$s_1(t-1) \geq 1 - \epsilon_1. \tag{A.3}$$

Since $\sigma(\cdot)$ is a monotonically increasing function, we can lower bound $s_1(t)$ by substituting the minimal value for the expression inside the brackets in (A.1). In this case, (A.1), (A.2), and (A.3) yield

$$s_1(t) \geq \sigma\big(15.2(1 - \epsilon_1) + 8.4 \cdot 0 + 0.2 \cdot 0.015 - 3 - 7.6\big)$$
$$= \sigma(-15.2\epsilon_1 + 4.6).$$

Thus, Condition 1 in Lemma A.1 holds if $\sigma(-15.2\epsilon_1 + 4.6) \geq 1 - \epsilon_1$. It is easy to verify that this indeed holds for any $\epsilon_1 \in (0.01, 0.219)$.

To analyze the second condition in Lemma A.1, suppose that $s_4(t-1) = s_4(t-2) = s_4(t-3) = 1$. It follows from (5.3) and (5.5) that $s_3(t-1) = \sigma(3.8) = 0.98$, and

$$s_2(t-1) \geq \sigma\left(-0.2 \cdot 0.8 + 4.5\sigma(3.8) + 1.5 - 4.7\right)$$
$$= 0.73.$$

Substituting these values in (A.1) yields

$$s_1(t) = \sigma\big(15.2s_1(t-1) + 8.4s_1(t-1) + 0.2 \cdot 0.98 + 3 - 7.6\big)$$
$$\geq \sigma\big(15.2s_1(t-1) + 8.4 \cdot 0.73 + 0.2 \cdot 0.98 + 3 - 7.6\big)$$
$$\geq \sigma(1.72),$$

where the last inequality follows from the fact that $s_1(t-1)$, being the output of a Logistic function, is non-negative. Thus, Condition 2 in Lemma A.1 will hold if $\sigma(1.72) \geq 1 - \epsilon_1$, or $\epsilon_1 \geq 0.152$.

To analyze Condition 3 of the lemma, suppose that $s_1(t-1) \leq \epsilon_2$. Then (A.1) yields

$$s_1(t) \leq \sigma\big(15.2\epsilon_2 + 8.4s_2(t-1) + 0.2s_3(t-1) + 3s_4(t-1) - 7.6\big).$$

We can upper bound this by substituting the maximal values for the expression inside the brackets. Note, however, that Condition 3 does not apply when $s_4(t-1) = s_4(t-2) = s_4(t-3) = 1$ (as this case is covered by Condition 2). Under this constraint, applying (A.2) yields

$$s_1(t) \leq \sigma\big(15.2\epsilon_2 + 8.4 \cdot 0.8 + 0.2 \cdot 0.98 - 3 - 7.6\big)$$
$$= \sigma(15.2\epsilon_2 - 3.684).$$

Thus, Condition 3 of Lemma A.1 will hold if $\sigma(15.2\epsilon_2 - 3.684) \leq \epsilon_2$, and it is easy to verify that this indeed holds for any $\epsilon_2 \in (0.06, 0.09)$.

Summarizing, for $\epsilon_1 \in [0.152, 0.219)$ and $\epsilon_2 \in (0.06, 0.09)$, the trained RNN satisfies all the conditions of Lemma A.1. This completes the proof of Proposition 5.6. $\square$

Proof of Proposition 6.1

The proof is similar to the proof of Proposition 5.6, namely, we show that the RNN defined by (5.4), (6.4), and (6.5) indeed satisfies the three conditions in Lemma A.1. Note that using (6.5) yields

$$s_1(t) = \sigma\big(5.04\alpha s_1(t-1) + 2\alpha s_2(t-1) + 2\alpha s_3(t-1)$$
$$+ 0.76\alpha s_4(t-1) - 3.52\alpha\big), \tag{A.4}$$

whereas substituting (5.4) in (6.1) and (6.2) yields

$$s_4(t) \in \{-1, 1\}, \quad \text{and} \quad s_2(t), s_3(t) \in \{0.12, 0.88\}. \tag{A.5}$$

Suppose that
$$s_1(t-1) \geq 1 - \epsilon_1. \tag{A.6}$$

Since $\sigma(\cdot)$ is a monotonically increasing function, we can lower bound $s_1(t)$ by substituting the minimal value for the expression inside the brackets in (A.4). In this case, (A.4), (A.5), and (A.6) yield

$$s_1(t) \geq \sigma\big(5.04\alpha(1 - \epsilon_1) + 0.24\alpha + 0.24\alpha - 0.76\alpha - 3.52\alpha\big)$$
$$= \sigma(-5.04\alpha\epsilon_1 + 1.24\alpha).$$

Thus, Condition 1 in Lemma A.1 holds if $\sigma(-5.04\alpha\epsilon_1 + 1.24\alpha) \geq 1 - \epsilon_1$. It is easy to verify that this indeed holds for any $\alpha \geq 5.7$, with $\epsilon_1 = 0.036$.

To analyze the second condition in Lemma A.1, suppose that $s_4(t-1) = s_4(t-2) = s_4(t-3) = 1$. It follows from (6.4) and (6.5) that $s_3(t-1) = \sigma(2) = 0.88$, and $s_2(t-1) = \sigma(5.26\sigma(2) - 2.62) = 0.88$. Substituting these values in (A.4) yields

$$s_1(t) = \sigma\big(5.04\alpha s_1(t-1) + 1.76\alpha + 1.76\alpha + 0.76\alpha - 3.52\alpha\big)$$
$$\geq \sigma(0.76\alpha),$$

where the inequality follows from the fact that $s_1(t-1)$ is non-negative. Thus, Condition 2 in Lemma A.1 will hold if $\sigma(0.76\alpha) \geq 1 - \epsilon_1$, and it is easy to verify that this indeed holds for any $\alpha \geq 5.7$, with $\epsilon_1 = 0.036$.

To analyze Condition 3 in Lemma A.1, suppose that $s_1(t-1) \leq \epsilon_2$. Then (A.4) yields

$$s_1(t) \leq \sigma\big(5.04\alpha\epsilon_2 + 2\alpha s_2(t-1) + 2\alpha s_3(t-1) + 0.76\alpha s_4(t-1) - 3.52\alpha\big).$$

We can upper bound this by substituting the maximal values for the expression inside the brackets. Note, however, that Condition 3 does not apply when $s_4(t-1) = s_4(t-2) = s_4(t-3) = 1$ (as this case is covered by Condition 2). Under this constraint, applying (A.5) yields

$$s_1(t) \leq \sigma\big(5.04\alpha\epsilon_2 + 1.76\alpha + 1.76\alpha - 0.76\alpha - 3.52\alpha\big)$$
$$= \sigma(5.04\alpha\epsilon_2 - 0.76\alpha).$$

Thus, Condition 3 of Lemma A.1 will hold if $\sigma(5.04\alpha\epsilon_2 - 0.76\alpha) \le \epsilon_2$ and it is easy to verify that this indeed holds for any $\alpha \ge 5.7$, with $\epsilon_2 = 0.036$.

Summarizing, for $\alpha \ge 5.7$ the designed RNN satisfies all the conditions of Lemma A.1 for the specific values $\epsilon_1 = \epsilon_2 = 0.036$. This completes the proof of Proposition 6.1. $\square$

B Details of the LED Recognition Network

The 24-6-10 ANN was trained using MATLAB's Neural Network Toolbox. Parameter values were initialized using the "init" command with "net.layers{i}.initFcn" set to "initnw" (implementing the Nguyen-Widrow algorithm [121]). Training was performed using the "trainlm" command (Levenberg-Marquardt backprop), with "net.performFcn" set to "msereg" (that is, using the regularization factor $\sum w_{ij}^2$ [69]).

The parameters of the trained ANN are as follows.[1] The weights from the inputs to the hidden neurons are:

$$W = \begin{pmatrix} 0.23 & -0.04 & 0.45 & 0.30 & -0.17 & -0.52 & -0.14 & \dots \\ -1.31 & -0.25 & -0.06 & 0.77 & 0.70 & 0.73 & 1.07 & \dots \\ -1.09 & -2.05 & -1.86 & 1.58 & 0.60 & -0.15 & -0.63 & \dots \\ 2.99 & 0.59 & -0.17 & 0.40 & -0.79 & 1.08 & -2.50 & \dots \\ -0.57 & -2.02 & -0.25 & -0.65 & -0.09 & 2.08 & 2.90 & \dots \\ -0.49 & 0.89 & 0.02 & -0.44 & -0.62 & -1.65 & 0.55 & \dots \end{pmatrix}.$$

The values w_{ij}, $j \in [8:24]$, $i \in [1:6]$, are omitted since they satisfy $|w_{ij}| \leq 8.3E - 5$. The hidden neurons biases are

$$\mathbf{b} = (0.33, -0.59, 1.63, -2.20, -1.90, 1.59)^T.$$

The weights from the hidden to the output neurons are:

$$C = \begin{pmatrix} -0.43 & -0.32 & 0.62 & 0.95 & 0.02 & -0.38 & -0.89 & 0.1 & 0.07 & 0.46 \\ -0.22 & -0.69 & -0.07 & 0.32 & 0.05 & 0.05 & 0.43 & -0.59 & 0.59 & 0.13 \\ -0.43 & 0.24 & 0.25 & 0.13 & 0.12 & 0.21 & 0.21 & -0.04 & -0.42 & -0.26 \\ -0.38 & -0.43 & -0.31 & 0.22 & 0.01 & 0.57 & 0.07 & 0.12 & -0.23 & 0.35 \\ 0.34 & -0.14 & -0.22 & 0.85 & -0.49 & 0.28 & -0.24 & -0.49 & -0.17 & 0.28 \\ 0.28 & -0.18 & 0.27 & -0.28 & 0.2 & 0.47 & -0.26 & -0.66 & -0.27 & 0.44 \end{pmatrix},$$

and the output neurons biases are:

$$\mathbf{\Phi} = -(0.74, 0.78, 1.13, 0.92, 0.89, 0.71, 0.45, 0.68, 0.74, 0.96)^T.$$

---

[1] All the numerical values were rounded to two decimal digits, without affecting the classification accuracy.

References

1. Aha, D.W., Kibler, D., Albert, M.K.: Instance-based learning algorithms. Machine Learning 6, 37–66 (1991)
2. Alquezar, R., Sanfeliu, A.: An algebraic framework to represent finite state machines in single-layer recurrent neural networks. Neural Computation 7, 931–949 (1995)
3. Andersen, H.C., Lotfi, A., Westphal, L.C.: Comments on Functional equivalence between radial basis function networks and fuzzy inference systems. IEEE Trans. Neural Networks 9, 1529–1531 (1998)
4. Andrews, R., Diederich, J., Tickle, A.B.: Survey and critique of techniques for extracting rules from trained artificial neural networks. Knowledge-Based Systems 8, 373–389 (1995)
5. Angeline, P.J., Saunders, G.M., Pollack, J.B.: An evolutionary algorithm that constructs recurrent neural networks. Neural Computation 5, 54–65 (1994)
6. Asuncion, A., Newman, D.: UCI machine learning repository (2007), http://www.ics.uci.edu/~mlearn/MLRepository.html
7. Ayoubi, M., Isermann, R.: Neuro-fuzzy systems for diagnosis. Fuzzy Sets and Systems 89, 289–307 (1997)
8. Banerjee, M., Mitra, S., Pal, S.K.: Rough fuzzy MLP: Knowledge encoding and classification. IEEE Trans. Neural Networks 9, 1203–1216 (1998)
9. Barakat, N., Diedrich, J.: Eclectic rule-extraction from support vector machines. Int. J. Computational Intelligence 2, 59–62 (2005)
10. Baranyi, P., Yam, Y.: Fuzzy rule base reduction. In: Ruan, D., Kerre, E.E. (eds.) Fuzzy IF-THEN Rules in Computational Intelligence: Theory and Applications, pp. 135–160. Kluwer, Dordrecht (2000)
11. Bellman, R.: Dynamic Programming. Princeton University Press, Princeton (1957)
12. Benitez, J.M., Castro, J.L., Requena, I.: Are artificial neural networks black boxes?. IEEE Trans. Neural Networks 8, 1156–1164 (1997)
13. Bernadó, E., Llorà, X., Garrell, J.M.: XCS and GALE: A comparative study of two learning classifier systems on data mining. In: Lanzi, P.L., Stolzmann, W., Wilson, S.W. (eds.) IWLCS 2001. LNCS (LNAI), vol. 2321, pp. 115–132. Springer, Heidelberg (2002)
14. Bishop, C.M.: Neural Networks for Pattern Recognition. Oxford University Press, Oxford (1995)

90   References

[15] Blair, A.D., Pollack, J.B.: Analysis of dynamical recognizers. Neural Computation 9, 1127–1142 (1997)

[16] Blanco, A., Delgado, M., Pegalajar, M.C.: Fuzzy automaton induction using neural network. Int. J. Approximate Reasoning 27, 1–26 (2001)

[17] Boden, M., Wiles, J., Tonkes, B., Blair, A.: Learning to predict a context-free language: analysis of dynamics in recurrent hidden units. In: Proc. 9th Int. Conf. Artificial Neural Networks (ICANN 1999), pp. 359–364 (1999)

[18] Boger, Z., Guterman, H.: Knowledge extraction from artificial neural network models. In: Proc. IEEE Int. Conf. Systems, Man and Cybernetics (SMC 1997), pp. 3030–3035 (1997)

[19] Bose, B.K.: Expert system, fuzzy logic, and neural network applications in power electronics and motion control. Proc. IEEE 82, 1303–1323 (1994)

[20] Breiman, L., Friedman, J., Olshen, R., Stone, C.: Classification and Regression Trees, pp. 18–55. Chapman & Hall, Boca Raton (1984)

[21] Buchanan, B.G., Sutherland, G.L., Feigenbaum, E.A.: Heuristic DENDRAL: A program for generating explanatory hypotheses in organic chemistry. In: Meltzer, B., Michie, D. (eds.) Machine Intelligence, vol. 4, pp. 209–254. Edinburgh University Press (1969)

[22] Burges, C.J.C.: A tutorial on support vector machines for pattern recognition. Data Mining and Knowledge Discovery 2, 121–167 (2000)

[23] Castro, J.L., Flores-Hidalgo, L.D., Mantas, C.J., Puche, J.M.: Extraction of fuzzy rules from support vector machines. Fuzzy Sets and Systems 158, 2057–2077 (2007)

[24] Castro, J.L., Mantas, C.J., Benitez, J.M.: Interpretation of artificial neural networks by means of fuzzy rules. IEEE Trans. Neural Networks 13, 101–116 (2002)

[25] Chomsky, N.: Three models for the description of language. IRE Trans. Information Theory IT-2, 113–124 (1956)

[26] Chomsky, N., Schützenberger, M.P.: The algebraic theory of context-free languages. In: Bradford, P., Hirschberg, D. (eds.) Computer Programming and Formal Systems, pp. 118–161. North-Holland, Amsterdam (1963)

[27] Cloete, I.: Knowledge-based neurocomputing: Past, present, and future. In: Cloete, I., Zurada, J.M. (eds.) Knowledge-Based Neurocomputing, pp. 1–26. MIT Press, Cambridge (2000)

[28] Cloete, I.: VL_1ANN: Transformation of rules to artificial neural networks. In: Cloete, I., Zurada, J.M. (eds.) Knowledge-Based Neurocomputing, pp. 207–216. MIT Press, Cambridge (2000)

[29] Cloete, I., Zurada, J.M. (eds.): Knowledge-Based Neurocomputing. MIT Press, Cambridge (2000)

[30] Craven, M.W.: Extracting comprehensible models from trained neural networks. Ph.D. dissertation, Dept. of Computer Science, University of Wisconsin–Madison (1996)

[31] Ruan, D.: Online experiments of controlling nuclear reactor power with fuzzy logic. In: Proc. IEEE Int. Conf. Fuzzy Systems (FUZZ-IEEE 1999), pp. 1712–1717 (1999)

[32] Darbari, A.: Rule extraction from trained ANN: A survey. Institute of Artificial intelligence, Dept. of Computer Science, Dresden Technology University, Tech. Rep. (2000)

[33] Dubois, D., Nguyen, H.T., Prade, H., Sugeno, M.: Introduction: The real contribution of fuzzy systems. In: Nguyen, H.T., Sugeno, M. (eds.) Fuzzy Systems: Modeling and Control, pp. 1–17. Kluwer, Dordrecht (1998)

[34] Duch, W.: Therapeutic implications of computer models of brain activity for Alzheimer's disease. J. Medical Informatics Technologies 5, 27–34 (2000)

[35] Duch, W., Adamczak, R., Crabczewski, K., Ishikawa, M., Ueda, H.: Extraction of crisp logical rules using constrained backpropagation networks-comparison of two new approaches. In: Proc. European Symp. Artificial Neural Networks (ESANN 1997), pp. 109–114 (1997)

[36] Duch, W., Adamczak, R., Grabczewski, K.: Extraction of logical rules from neural networks. Neural Processing Letters 7, 211–219 (1998)

[37] Duch, W., Adamczak, R., Grabczewski, K., Zal, G.: Hybrid neural-global minimization method of logical rule extraction. J. Advanced Computational Intelligence 3, 348–356 (1999)

[38] Etchells, T., Lisboa, P.J.G.: Orthogonal search-based rule extraction (OSRE) for trained neural networks: A practical and efficient approach. IEEE Trans. Neural Networks 17, 374–384 (2006)

[39] Fisher, R.A.: The use of multiple measurements in taxonomic problems. Annual Eugenics 7, 179–188 (1936)

[40] Fodor, J.A., Pylyshyn, Z.W.: Connectionism and cognitive architecture: A critical analysis. In: Pinker, S., Mehler, J. (eds.) Connections and Symbols, pp. 3–72. MIT Press, Cambridge (1988)

[41] Frasconi, P., Gori, M., Maggini, M., Soda, G.: Representation of finite-state automata in recurrent radial basis function networks. Machine Learning 23, 5–32 (1996)

[42] Fu, L.-M.: Knowledge-based connectionism for revising domain theories. IEEE Trans. Systems, Man and Cybernetics 23, 173–182 (1993)

[43] Fu, L.-M.: Neural Networks in Computer Science. McGraw-Hill, New York (1994)

[44] Fu, L.-M.: Learning capacity and sample complexity on expert networks. IEEE Trans. Neural Networks 7, 1517–1520 (1996)

[45] Fu, L.-M., Fu, L.-C.: Mapping rule-based systems into neural architecture. Knowledge-Based Systems 3, 48–56 (1990)

[46] Fukumi, M., Mitsukura, Y., Akamatsu, N.: Knowledge incorporation and rule extraction in neural networks. In: Dorffner, G., Bischof, H., Hornik, K. (eds.) ICANN 2001. LNCS, vol. 2130, pp. 1248–1254. Springer, Heidelberg (2001)

[47] Gallant, S.I.: Connectionist expert systems. Communications of the ACM 31, 152–169 (1988)

[48] Ganchev, T., Tasoulis, D.K., Vrahatis, M.N., Fakotakis, N.: Generalized locally recurrent probabilistic neural networks for text-independent speaker verification. In: Proc. IEEE Int. Conf. Acoustics, Speech, and Signal Processing (ICASSP 2004), vol. 1, pp. 41–44 (2004)

[49] Giles, C.L., Lawrence, S., Tsoi, A.C.: Noisy time series prediction using recurrent neural networks and grammatical inference. Machine Learning 44, 161–183 (2001)

[50] Giles, C.L., Miller, C.B., Chen, D., Chen, H.H., Sun, G.Z., Lee, Y.C.: Learning and extracting finite state automata with second-order recurrent neural networks. Neural Computation 4, 393–405 (1992)

[51] Giles, C.L., Omlin, C.W., Thornber, K.K.: Equivalence in knowledge representation: automata, recurrent neural networks, and dynamical fuzzy systems. In: Medsker, L.R., Jain, L.C. (eds.) Recurrent Neural Networks: Design and Applications. CRC Press, Boca Raton (1999)

[52] Gori, M., Maggini, M., Martinelli, E., Soda, G.: Inductive inference from noisy examples using the hybrid finite state filter. IEEE Trans. Neural Networks 9, 571–575 (1998)

92   References

[53] Goudreau, M.W., Giles, C.L., Chakradhar, S.T., Chen, D.: First-order versus second-order single-layer recurrent neural networks. IEEE Trans. Neural Networks 5, 511–513 (1994)

[54] Guillaume, S.: Designing fuzzy inference systems from data: An interpretability-oriented review. IEEE Trans. Fuzzy Systems 9, 426–443 (2001)

[55] Gupta, A., Lam, S.M.: Generalized analytic rule extraction for feedforward neural networks. IEEE Trans. Knowledge and Data Engineering 11, 985–992 (1999)

[56] Hani, M.K., Nor, S.M., Hussein, S., Elfadil, N.: Machine learning: The automation of knowledge acquisition using Kohonen self-organising map neural network. Malysian J. Computer Science 14, 68–82 (2001)

[57] Hasselmo, M.E.: Runaway synaptic modification in models of cortex: Implications for Alzheimer's disease. Neural Networks 7, 13–40 (1991)

[58] Hasselmo, M.E.: A computational model of the progression of Alzheimer's disease. MD Computing 14, 181–191 (1997)

[59] Hassoun, M.H.: Fundamentals of Artificial Neural Networks. MIT Press, Cambridge (1995)

[60] Hebb, D.O.: The Organization of Behavior. Wiley, Chichester (1949)

[61] Holena, M.: Piecewise-linear neural networks and their relationship to rule extraction from data. Neural Computation 18, 2813–2853 (2006)

[62] Hopcroft, J.E., Motwani, R., Ullman, J.D.: Introduction to Automata Theory, Languages and Computation, 2nd edn. Addison-Wesley, Reading (2001)

[63] Horn, D., Levy, N., Ruppin, E.: Neuronal-based synaptic compensation: A computational study in Alzheimer's disease. Neural Computation 8, 1227–1243 (1996)

[64] Horn, D., Ruppin, E., Usher, M., Herrmann, M.: Neural network modeling of memory deterioration in Alzheimer's disease. Neural Computation 5, 736–749 (1993)

[65] Huang, S.H.: Dimensionality reduction in automatic knowledge acquisition: A simple greedy search approach. IEEE Trans. Knowledge and Data Engineering 15, 1364–1373 (2003)

[66] Huang, S.H., Xing, H.: Extract intelligible and concise fuzzy rules from neural networks. Fuzzy Sets and Systems 132, 233–243 (2002)

[67] Hunter, L., Klein, T.: Finding relevant biomolecular features. In: Proc. 1st Int. Conf. Intelligent Systems for Molecular Biology, pp. 190–197 (1993)

[68] Intrator, O., Intrator, N.: Interpreting neural-network results: A simulation study. Computational Statistics and Data Analysis 37, 373–393 (2001)

[69] Ishikawa, M.: Structural learning and rule discovery. In: Cloete, I., Zurada, J.M. (eds.) Knowledge-Based Neurocomputing, pp. 153–206. MIT Press, Cambridge (2000)

[70] Jackson, P.: Introduction to Expert Systems. Addison-Wesley, Reading (1986)

[71] Jacobsson, H.: Rule extraction from recurrent neural networks: A taxonomy and review. Neural Computation 17, 1223–1263 (2005)

[72] Jamshidi, M., Titly, A., Zadeh, L.A., Boverie, S. (eds.): Applications of Fuzzy Logic: Towards High Machine Intelligence Quotient Systems. Prentice Hall, Englewood Cliffs (1997)

[73] Jang, J.-S.R.: ANFIS: Adaptive-network-based fuzzy inference system. IEEE Trans. Systems, Man and Cybernetics 23, 665–685 (1993)

[74] Jang, J.-S.R., Sun, C.-T.: Functional equivalence between radial basis function networks and fuzzy inference systems. IEEE Trans. Neural Networks 4, 156–159 (1993)

[75] Jang, J.-S.R., Sun, C.-T., Mizutani, E.: Neuro-Fuzzy and Soft Computing: A Computational Approach to Learning and Machine Intelligence. Prentice-Hall, Englewood Cliffs (1997)

[76] Jassem, W., Grygiel, W.: Off-line classification of Polish vowel spectra using artificial neural networks. J. Int. Phonetic Association 34, 37–52 (2004)

[77] Jin, Y.: Fuzzy modeling of high-dimensional systems: Complexity reduction and interpretability improvement. IEEE Trans. Fuzzy Systems 8, 212–221 (2000)

[78] Jin, Y., Von Seelen, W., Sendhoff, B.: On generating FC3 fuzzy rule systems from data using evolution strategies. IEEE Trans. Systems, Man and Cybernetics 29, 829–845 (1999)

[79] Kahneman, D., Tversky, A.: Prospect theory: An analysis of decision under risk. Econometrica 47, 263–292 (1979)

[80] Kandel, A., Langholz, G. (eds.): Hybrid Architectures for Intelligent Systems. CRC Press, Boca Raton (1992)

[81] Kasabov, N.K., Kim, J., Watts, M.J., Gray, A.R.: FuNN/2–a fuzzy neural network architecture for adaptive learning and knowledge acquisition. Information Sciences 101, 155–175 (1997)

[82] Kavzoglu, T., Mather, P.M.: The use of backpropagating artificial neural networks in land cover classification. Int. J. Remote Sensing 24, 4907–4938 (2003)

[83] Khan, E., Unal, F.: Recurrent fuzzy logic using neural networks. In: Furuhashi, T. (ed.) Advances in Fuzzy Logic, Neural Networks, and Genetic Algorithms, pp. 48–55. Springer, Heidelberg (1995)

[84] Kleene, S.C.: Representation of events in nerve nets and finite automata. In: Shannon, C.E., McCarthy, J. (eds.) Automata Studies, ser. Annals of Mathematics Studies, vol. 34, pp. 3–41. Princeton University Press, Princeton (1956)

[85] Kohavi, R.: A study of cross-validation and bootstrap for accuracy estimation and model selection. In: Proc. 14th Int. Joint Conf. on Artificial Intelligence (IJCAI 1995), pp. 1137–1145 (1995)

[86] Kolen, J.F.: Fool's gold: Extracting finite state machines from recurrent network dynamics. In: Cowan, J.D., Tesauro, G., Alspector, J. (eds.) Advances in Neural Information Processing Systems 6, pp. 501–508. Morgan Kaufmann, San Francisco (1994)

[87] Kolen, J.F., Pollack, J.B.: The observer's paradox: Apparent computational complexity in physical systems. J. Experimental and Theoretical Artificial Intelligence 7, 253–277 (1995)

[88] Kolman, E., Margaliot, M.: Extracting symbolic knowledge from recurrent neural networks–a fuzzy logic approach. Fuzzy Sets and Systems (to appear), `www.eng.tau.ac.il/~michaelm`

[89] Kolman, E., Margaliot, M.: A new approach to knowledge-based design of recurrent neural networks. IEEE Trans. Neural Networks 19(8), 1389–1401 (2008)

[90] Kolman, E., Margaliot, M.: Are artificial neural networks white boxes?. IEEE Trans. Neural Networks 16, 844–852 (2005)

[91] Kolman, E., Margaliot, M.: Knowledge extraction from neural networks using the all-permutations fuzzy rule base: The LED display recognition problem. IEEE Trans. Neural Networks 18, 925–931 (2007)

[92] Kreinovich, V., Langrand, C., Nguyen, H.T.: A statistical analysis for rule base reduction. In: Proc. 2nd Int. Conf. Intelligent Technologies (InTech 2001), pp. 47–52 (2001)

[93] Kuan, C.-M., Liu, T.: Forecasting exchange rates using feedforward and recurrent neural networks. J. Applied Econometrics 10, 347–364 (1995)

94   References

[94] Lang, K.J.: Random DFA's can be approximately learned from sparse uniform examples. In: Proc. 5th ACM Workshop Computational Learning Theory, pp. 45–52 (1992)

[95] LeCun, Y.: Learning processes in an asymmetric threshold network. In: Bienenstock, E., Fogelman-Soulie, F., Weisbuch, G. (eds.) Disordered Systems and Biological Organization, pp. 233–340. Springer, Heidelberg (1986)

[96] Lee, S.-W., Song, H.-H.: A new recurrent neural network architecture for visual pattern recognition. IEEE Trans. Neural Networks 8, 331–340 (1997)

[97] Leng, G., McGinnity, T.M., Prasad, G.: An approach for on-line extraction of fuzzy rules using a self-organizing fuzzy neural network. Fuzzy Sets and Systems 150, 211–243 (2005)

[98] Lin, C.-T., Lin, C.S.G.: Neural-network-based fuzzy logic control and decision system. IEEE Trans. Computers 40, 1320–1336 (1991)

[99] Linggard, R., Myers, D.J., Nightingale, C. (eds.): Neural Networks for Vision, Speech, and Natural Language. Chapman & Hall, Boca Raton (1992)

[100] Llora, X.: Genetic based machine learning using fine-grained parallelism for data mining. Ph.D. dissertation, Enginyeria i Arquitectura La Salle, Ramon Llull University (2002)

[101] Llora, X., Goldberg, E.: Bounding the effect of noise in multiobjective learning classifier systems. Evolutionary Computation 11, 279–298 (2003)

[102] Maire, F.: Rule-extraction by backpropagation of polyhedra. Neural Networks 12, 717–725 (1999)

[103] Mamdani, E.H.: Applications of fuzzy algorithms for simple dynamic plant. Proc. IEE 121, 1585–1588 (1974)

[104] Mamdani, E.H., Assilian, S.: An experiment in linguistic synthesis with a fuzzy logic controller. Int. J. Man-Machine Studies 7, 1–13 (1975)

[105] Manolios, P., Fanelly, R.: First order recurrent neural networks and deterministic finite state automata. Neural Computation 6, 1155–1173 (1994)

[106] Margaliot, M., Langholz, G.: Hyperbolic optimal control and fuzzy control. IEEE Trans. Systems, Man and Cybernetics 29, 1–10 (1999)

[107] Margaliot, M., Langholz, G.: New Approaches to Fuzzy Modeling and Control - Design and Analysis. World Scientific, Singapore (2000)

[108] McCarthy, J.: What is artificial intelligence? (2007), http://www-formal.stanford.edu/jmc/whatisai.pdf

[109] McCarthy, J., Minsky, M.L., Rochester, N., Shannon, C.E.: A proposal for the Dartmouth summer research project on artificial intelligence (1956), http://www-formal.stanford.edu/jmc/history/dartmouth/dartmouth.html

[110] McCulloch, W.S., Pitts, W.H.: A logical calculus of the ideas imminent in nervous activity. Bulletin of Mathematical Biophysics 5, 115–137 (1943)

[111] McGarry, K., Wermter, S., MacIntyre, J.: Hybrid neural systems: From simple coupling to fully integrated neural networks. Neural Computing Surveys 2, 62–93 (1999)

[112] Medsker, L.R., Jain, L.C. (eds.): Recurrent Neural Networks: Design and Applications. CRC Press, Boca Raton (1999)

[113] Miller, C.B., Giles, C.L.: Experimental comparison of the effect of order in recurrent neural networks. Int. J. Pattern Recognition and Artificial Intelligence 7, 849–872 (1993)

[114] Miller, D.A., Zurada, J.M.: A dynamical system perspective of structural learning with forgetting. IEEE Trans. Neural Networks 9, 508–515 (1998)

[115] Miller, W.T., Sutton, R.S., Werbos, P.J. (eds.): Neural Networks for Control. MIT Press, Cambridge (1990)

[116] Minsky, M.L., Papert, S.: Perceptrons: An Introduction to Computational Geometry. MIT Press, Cambridge (1969)

[117] Mitra, S., De, R.K., Pal, S.K.: Knowledge-based fuzzy MLP for classification and rule generation. IEEE Trans. Neural Networks 8, 1338–1350 (1997)

[118] Mitra, S., Hayashi, Y.: Neuro-fuzzy rule generation: Survey in soft computing framework. IEEE Trans. Neural Networks 11, 748–768 (2000)

[119] Mitra, S., Pal, S.K.: Fuzzy multi-layer perceptron, inferencing and rule generation. IEEE Trans. Neural Networks 6, 51–63 (1995)

[120] Murray, A. (ed.): Applications of Neural Networks. Kluwer, Dordrecht (1995)

[121] Nguyen, D., Widrow, B.: Improving the learning speed of 2-layer neural networks by choosing initial values of the adaptive weights. In: Proc. 3rd Int. Joint Conf. Neural Networks, pp. 21–26 (1990)

[122] Novak, V.: Are fuzzy sets a reasonable tool for modeling vague phenomena? Fuzzy Sets and Systems 156, 341–348 (2005)

[123] Nunez, H., Angulo, C., Catala, A.: Rule extraction from support vector machines. In: Proc. 10th European Symp. Artificial Neural Networks, pp. 107–112 (2002)

[124] Omlin, C.W., Giles, C.L.: Extraction and insertion of symbolic information in recurrent neural networks. In: Honavar, V., Uhr, L. (eds.) Artificial Intelligence and Neural Networks: Steps Toward Principled Integration, pp. 271–299. Academic Press, London (1994)

[125] Omlin, C.W., Giles, C.L.: Constructing deterministic finite-state automata in recurrent neural networks. J. ACM 43, 937–972 (1996)

[126] Omlin, C.W., Giles, C.L.: Extraction of rules from discrete-time recurrent neural networks. Neural Networks 9, 41–52 (1996)

[127] Omlin, C.W., Giles, C.L.: Symbolic knowledge representation in recurrent neural networks: Insights from theoretical models of computation. In: Cloete, I., Zurada, J.M. (eds.) Knowledge-Based Neurocomputing, pp. 63–115. MIT Press, Cambridge (2000)

[128] Omlin, C.W., Thornber, K.K., Giles, C.L.: Fuzzy finite-state automata can be deterministically encoded into recurrent neural networks. IEEE Trans. Fuzzy Systems 6, 76–89 (1998)

[129] Paiva, R.P., Dourado, A.: Interpretability and learning in neuro-fuzzy systems. Fuzzy Sets and Systems 147, 17–38 (2004)

[130] Parker, D.B.: Learning logic. Center for Computational Research in Economics and Management Science. MIT, Tech. Rep (1985)

[131] Pedrycz, W.: Fuzzy Control and Fuzzy Systems, 2nd edn. Wiley, Chichester (1993)

[132] Pedrycz, W., Kandel, A., Zhang, Y.-Q.: Neurofuzzy systems. In: Nguyen, H.T., Sugeno, M. (eds.) Fuzzy Systems: Modeling and Control, pp. 311–380. Kluwer, Dordrecht (1998)

[133] Pedrycz, W., Rocha, A.F.: Fuzzy-set based models of neurons and knowledge-based networks. IEEE Trans. Fuzzy Systems 1, 254–266 (1993)

[134] Pollack, J.B.: Cascaded back-propagation on dynamic connectionist networks. In: Proc. 9th Annual Conf. Cognitive Science Society, pp. 391–404 (1987)

[135] Pollack, J.B.: The induction of dynamical recognizers. Machine Learning 7, 227–252 (1991)

[136] Rashkovsky, I., Margaliot, M.: Nicholson's blowflies revisited: A fuzzy modeling approach. Fuzzy Sets and Systems 158, 1083–1096 (2007)

96   References

[137] Reggia, J.A., Ruppin, E., Glanzman, D.L. (eds.): Neural Modeling of Brain and Cognitive Disorders. World Scientific, Singapore (1996)

[138] Robinson, T., Hochberg, M., Renals, S.: The use of recurrent networks in continuous speech recognition. In: Lee, C.H., Soong, F.K., Paliwal, K.K. (eds.) Automatic Speech and Speaker Recognition: Advanced Topics, pp. 233–258. Kluwer, Dordrecht (1996)

[139] Rodriguez, P.: Simple recurrent networks learn context-free and context-sensitive languages by counting. Neural Computation 13, 2093–2118 (2001)

[140] Rodriguez, P., Wiles, J., Elman, J.L.: A recurrent neural network that learns to count. Connection Science 11, 5–40 (1999)

[141] Rosenblatt, F.: The perceptron: A perceiving and recognizing automaton. Project PARA, Cornell Aeronautical Laboratory, Tech. Rep. (1957)

[142] Rowley, H.A., Baluja, S., Kanade, T.: Neural network-based face detection. IEEE Trans. Pattern Analysis and Machine Intelligence 20, 23–38 (1998)

[143] Rozin, V., Margaliot, M.: The fuzzy ant. IEEE Computational Intelligence Magazine 2, 18–28 (2007)

[144] Rumelhart, D.E., Hinton, G.E., Williams, R.J.: Learning internal representation by error propagation. In: Rumelhart, D.E., McClelland, J.L. (eds.) Parallel Distributed Processing: Explorations in the Microstructure of Cognition, pp. 318–362. MIT Press, Cambridge (1986)

[145] Rumelhart, D.E., McClelland, J.L. (eds.): Parallel Distributed Processing, vol. 2. MIT Press, Cambridge (1986)

[146] Ruppin, E., Reggia, J.: A neural model of memory impairment in diffuse cerebral atrophy. British J. Psychiatry 166, 19–28 (1995)

[147] Russell, S.J., Norvig, P.: Artificial Intelligence: A Modern Approach, 2nd edn. Prentice Hall, Englewood Cliffs (2003)

[148] Saito, K., Nakano, R.: Medical diagnostic expert system based on PDP model. In: Proc. 1st IEEE Int. Conf. Neural Networks, pp. 255–262 (1988)

[149] Saito, K., Nakano, R.: Rule extraction from facts and neural networks. In: Proc. 2nd Int. Conf. Neural Networks (INNC 1990), pp. 379–382 (1990)

[150] Saito, K., Nakano, R.: Second-order learning algorithm with squared penalty term. Neural Computation 12, 709–729 (2000)

[151] Samuel, A.L.: Some studies in machine learning using the game of checkers. IBM J. Research and Development 3, 211–229 (1959)

[152] Saund, E.: Dimensionality-reduction using connectionist networks. IEEE Trans. Pattern Analysis and Machine Intelligence 11, 304–314 (1989)

[153] Schmitz, G., Aldrich, C., Gouws, F.: Extraction of decision trees from artificial neural networks. In: Cloete, I., Zurada, J.M. (eds.) Knowledge-Based Neurocomputing, pp. 369–401. MIT Press, Cambridge (2000)

[154] Sejnowski, T.J., Rosenberg, C.R.: Parallel networks that learn to pronounce English text. Complex Systems 1, 145–168 (1987)

[155] Sestito, S., Dillon, T.: Knowledge acquisition of conjunctive rules using multilayered neural networks. Int. J. Intelligent Systems 8, 779–805 (1993)

[156] Setiono, R.: Extracting rules from neural networks by pruning and hidden-unit splitting. Neural Computation 9, 205–225 (1997)

[157] Setiono, R., Tanaka, M.: Neural network rule extraction and the LED display recognition problem. In: IEEE Trans. Neural Networks (to appear)

[158] Shawe-Taylor, J., Cristianini, N.: Kernel Methods for Pattern Analysis. Cambridge University Press, Cambridge (2004)

[159] Shortliffe, E.H.: Computer-Based Medical Consultations: MYCIN. Elsevier, Amsterdam (1976)

[160] Siegelmann, H.T.: Foundations of recurrent neural networks. Ph.D. dissertation, Dept. of Computer Science, Rutgers (1993)

[161] Smolensky, P.: On the proper treatment of connectionism. Behavioral and Brain Science 11, 1–23 (1988)

[162] Snyders, S., Omlin, C.W.: What inductive bias gives good neural network training performance? In: Proc. IEEE-INNS-ENNS Int. Joint Conf. Neural Networks, pp. 445–450 (2000)

[163] Snyders, S., Omlin, C.W.: Inductive bias in recurrent neural networks. In: Proc. 6th Int. Work-Conf. Artificial and Natural Neural Networks, pp. 339–346 (2001)

[164] Sousa, J.W.C., Kaymak, U.: Fuzzy Decision Making in Modeling and Control. World Scientific, Singapore (2002)

[165] Terano, T., Asai, K., Sugeno, M.: Applied Fuzzy Systems. AP Professional (1994)

[166] Tickle, A.B., Andrews, R., Golea, M., Diederich, J.: The truth will come to light: Directions and challenges in extracting the knowledge embedded within trained artificial neural networks. IEEE Trans. Neural Networks 9, 1057–1068 (1998)

[167] Tickle, A.B., Golea, M., Hayward, R., Diederich, J.: The truth is in there: Current issues in extracting rules from trained feedforward artificial neural networks. In: Proc. 4th IEEE Int. Conf. Neural Networks, pp. 2530–2534 (1997)

[168] Tickle, A.B., Orlowski, M., Diederich, J.: DEDEC: A methodology for extracting rules from trained artificial neural networks. In: Andrews, R., Diederich, J. (eds.) Rules and Networks, pp. 90–102. QUT Publication (1996)

[169] Tino, P., Koteles, M.: Extracting finite-state representation from recurrent neural networks trained on chaotic symbolic sequences. Neural Computation 10, 284–302 (1999)

[170] Tino, P., Vojtec, V.: Extracting stochastic machines from recurrent neural networks trained on complex symbolic sequences. Neural Network World 8, 517–530 (1998)

[171] Tomita, M.: Dynamic construction of finite-state automata from examples using hill-climbing. In: Proc. 4th Ann. Conf. Cognitive Science, pp. 105–108 (1982)

[172] Towell, G.G., Shavlik, J.W.: Extracting refined rules from knowledge-based neural networks. Machine Learning 13, 71–101 (1993)

[173] Towell, G.G., Shavlik, J.W., Noordenier, M.: Refinement of approximate domain theories by knowledge based neural network. In: Proc. 8th Natl. Conf. Artificial Intelligence, pp. 861–866 (1990)

[174] Tron, E., Margaliot, M.: Mathematical modeling of observed natural behavior: A fuzzy logic approach. Fuzzy Sets and Systems 146, 437–450 (2004)

[175] Tron, E., Margaliot, M.: How does the Dendrocoleum lacteum orient to light? a fuzzy modeling approach. Fuzzy Sets and Systems 155, 236–251 (2005)

[176] Tsukimoto, H.: Extracting rules from trained neural networks. IEEE Trans. Neural Networks 11, 377–389 (2000)

[177] Tu, J.V.: Advantages and disadvantages of using artificial neural networks versus logistic regression for predicting medical outcomes. J. Clinical Epidemiology 49, 1225–1231 (1996)

[178] Watrous, R.L., Kuhn, G.M.: Induction of finite-state languages using second-order recurrent networks. Neural Computation 4, 406–414 (1992)

[179] Wei, Z., Wu, C., Chen, L.: Symbolism and connectionism of artificial intelligence. In: Proc. IEEE Asia-Pacific Conf. on Circuits and Systems (IEEE APCCAS), pp. 364–366 (2000)

98   References

[180] Werbos, P.J.: The Roots of Backpropagation: From Ordered Derivatives to Neural Networks and Political Forecasting. Wiley, Chichester (1994)

[181] Wiles, J., Elman, J.L.: Learning to count without a counter: A case study of dynamics and activation landscapes in recurrent networks. In: Proc. 17th Annual Conf. Cognitive Science Society, pp. 482–487 (1995)

[182] Williams, R.: Simple statistical gradient-following algorithms for connectionist reinforcement learning. Machine Learning 8, 229–256 (1992)

[183] Williams, R.J., Zipser, D.: A learning algorithm for continually running fully recurrent neural networks. Neural Computation 1, 270–280 (1989)

[184] Zadeh, L.A.: Fuzzy sets. Information and Control 8, 338–353 (1965)

[185] Zadeh, L.A.: Outline of a new approach to the analysis of complex systems and decision processes. IEEE Trans. Sytems, Man and Cybernatics 3, 28–44 (1973)

[186] Zadeh, L.A.: The concept of a linguistic variable and its application to approximate reasoning. Information Sciences 30, 199–249 (1975)

[187] Zadeh, L.A.: Fuzzy logic = computing with words. IEEE Trans. Fuzzy Systems 4, 103–111 (1996)

[188] Zeng, Z., Goodman, R.M., Smyth, P.: Learning finite state machines with self-clustering recurrent networks. Neural Computation 5, 976–990 (1993)

[189] Zhang, D., Bai, X.-L., Cai, K.-Y.: Extended neuro-fuzzy models of multilayer perceptrons. Fuzzy Sets and Systems 142, 221–242 (2004)

[190] Zhou, Z.-H.: Rule extraction: Using neural networks or for neural networks. J. Computer Science and Technology 19, 249–253 (2004)

[191] Zhou, Z.-H., Chen, S.-F., Chen, Z.-Q.: A statistics based approach for extracting priority rules from trained neural networks. In: Proc. 3rd Int. Joint Conf. Neural Networks (IJCNN 2000), pp. 401–406 (2000)

[192] Zurada, J.M.: Introduction to Artificial Neural Systems. West Publishing Company (1992)

Index

σ 8, 17
σ_L 18
Alzheimer's disease   80
ANFIS   5
ANNs, *see* Artificial Neural Networks
Artificial neural networks   2
  simplification   37

Back-propagation algorithm   2, 10
Benitez et al. model   7
Benitez, J.M.   7
Black-box   3, 5
Bootstrap   52

Chomsky, N.   50
Classifier
  linear   2
Cloete, I.   5, 10
COG, *see* Defuzzifier
Connectionism   1
Craven, M.W.   6
Cross-validation   52
Curse of dimensionality   77

DEDEC   6
Defuzzifier
  center of gravity (COG)   13, 18, 40
  mean of maxima (MOM)   40
DENDRAL   3
Deterministic finite-state automaton   9, 10, 51
DFA, *see* Deterministic Finite-state Automaton
Duch, W.   7

Expert systems   3

FARB   11, 13, 15
  and feedforward ANNs   21
  and first-order RNNs   26
  and second-order RNNs   33
  definition   15
Feedforward neural networks   10
FFA, *see* Fuzzy Finite-state Automaton
Formal grammar   50
  definition   50
  regular   50
Formal languages   50
  and RNNs   51
  context-free   50
  context-sensitive   50
  definition   50
  recursive   50
  regular   50
  Tomita's 4th grammar   51
Fu, L.-C.   10
Fu, L.-M.   10
Fuzzy-MLP   7
Fuzzy finite-state automaton   11
Fuzzy recognizer   71
Fuzzy rule-bases   3, 4
  simplification   37

Gallant, S.I.   10

Hard threshold unit   69
Hebb, D.O.   2
Hebbian learning   2
Holdout approach   52
Hopfield, J.J.   2

100 Index

interactive-or operator   8
Iris classification problem   41
Ishikawa, M.   7

Jang, J.-S.R.   7
Jang and Sun model   7

Kasabov, N.K.   10
KBANN   10
KBCNN   10
KBD, *see* Knowledge-based design
KE, *see* Knowledge extraction
Kleene closure operator   50
Knowledge-based design   10, 59
   direct approach   60
   modular approach   63
   of feedforward ANNs   10
   of recurrent ANNs   10
   orthogonal internal representation   10
Knowledge-based neurocomputing   5, 19
   hybrid   5
   translational   5
   unified   5
Knowledge extraction   5, 41
   decompositional   6
   eclectic   6
   from feedforward ANNs   6
   from recurrent ANNs   9
   pedagogical   6
Knowledge refinement   12, 81
Knowledge translator algorithm   6
KT algorithm, *see* Knowledge Translator
   Algorithm

L_4 language   51
   extended   69
Language recognition problem   50
Learned intermediaries doctrine   5
LED display recognition problem   44
Levenberg-Marquardt backprop
   algorithm   87

Mamdani, E.H.   4
McCarthy, J.   1
McCulloch, W.S.   2

Minsky, M.L.   2
Mitra, S.   7
MLP, *see* Multi-Layer Perceptron
MofN algorithm   6
MOM, *see* Defuzzifier
Multi-layer perceptron   2
MYCIN   3

Nguyen-Widrow algorithm   87

Pal, S.K.   7
Papert, S.   2
Perceptron   2
Pitts, W.   2
Pollack, J.B.   79

Real time recurrent learning   52
Recurrent neural networks   9
   knowledge-based design   10
Regularization   7, 78
RNNs, *see* Recurrent Neural Networks
Rosenblatt, F.   2
RTRL, *see* Real Time Recurrent Learning
Rule simplification   37, 46, 53, 55, 56
   sensitivity analysis   37

Samuel, L.A.   2
Schützenberger, M.P.   50
Shavlik, J.W.   10
Subset algorithm   6
Sun, C.-T.   7
Support vector machine   79
SVM, *see* Support Vector Machine
Symbolism   1

Tomita's 4th grammar   51
Towell, G.G.   10
TREPAN   6

White-box   4
Winner-takes-all approach   45

Xor   22
Xornot   24

Zadeh, L.A.   3, 4
   principle of incompatibility   3, 37